Grass Farmers

GRASS FARMERS

by

Allan Nation

Green Park Press
A division of Mississippi Valley Publishing, Corp.
Jackson, Mississippi

Dedicated to Dub Shoemaker

Library of Congress Catalog Card Number: 93-78111

ISBN: 0-9632460-1-1

Photographs by Allan Nation

Edited by Carolyn Nation

Cover Design by Heritage Graphics, Jackson, MS

Manufacturered in the United States of America
This book is printed on recycled paper.

Table of Contents

Foreword   7
1 Grassfed Beef in Virginia   11
2 Pasturing Poultry Pays   17
3 Buying the Farm--Fast   22
4 Grass Farm Looks Good as Potential
       Retirement Job   26
5 Balanced Forage Flow California
       Grass Farmer's Goal   29
6 Oil and Cattle No Longer Mix
       On South Texas'King Ranch   32
7 Relearning How to Farm   38
8 Cows, Clover, Compost Revive Cotton Country   41
9 Drought Proves Benefit of Perennial Pasture
       System   45
10 Upper Midwest Returning to Its Roots   50
11 God's Own Wintergrazing   55
12 Cows and Clover Seed Replace Soybeans on
       Mississippi Plantation   61
13 Back from the Brink   66
14 Grass Dairying in Texas   69
15 Climate Offers Dizzy Diversity   74
16 Organic Hogs on Pasture   80

17 Partnering in Alabama   83
18 Balanced Year Round Forage Aim of This Midwest
      Grass Farm   87
19 Working with Nature Key to Adams Ranch   91
20 A Vision of Grass Guides Anderson Ranch   95
21 Midwestern Ranchers Find Southern Grazing
      Connection Profitable   101
22 Managing for Quality Browse   105
23 Easy Does It...   107
24 Heifer Grazing Gives Grazier a Mid-winter
      Vacation   109
25 As Good As It Gets   111
26 Sheep Dairying Can Produce a Quality Life from
      Small Acreages   114
27 Mississippi Ranch Takes on Change in Big Bites   117
28 Parker Ranch Mixes Intensive and Extensive
      Grazing   123
29 Beef and Potatoes Profitable Mix in Northern
      Mexico   128
30 Polycultural Grazing in Arkansas   133
31 Way Down South...in Hawaii   140
32 Cow-calf Economics Shine with Low-input
      Program   143
33 Spring Lambing Ideal for Low-cost Pasture
      Program   146
34 Battening Down the Hatches   149
35 Snow Grazing, Canadian Style   153
36 Working with Nature Works   163
37 Work Fit for Man   166
Glossary   170
Index   178
Author's Bio   184
Ordering Information   185

Foreword

In 1977 I became the editor of **The Stockman Grass Farmer**. I had grown up on a commercial cattle ranch but had left to chase the bright lights. With a deep love for print journalism, I drifted into television advertising that paid better, but I was never really happy in the job. Getting into Ag journalism was like coming home. It allowed me to combine my love of writing with my heritage as a rancher's son.

At that time the magazine was called **The Stockman Farmer** and was your typical beef magazine. It was, as most advertising-dependent beef publications are, heavily oriented toward seedstock production and marketing. With my background in commercial cattle production, I was never comfortable with the "after tax" economics of the pre-Tax Reform seedstock business. I longed to cover fields of endeavor that produced real taxable income. In the late 1970's, the only group consistently making good money with beef cattle were the stocker graziers. I began to study and cover them in the magazine. Becoming aware of the skills needed to manipulate and maintain pasture quality, **I started to have a vague sense that the quality of the pasture was more important that the price of cattle.**

In 1979, we held our first Grazing Conference to explore the "state of the art" in stocker graziering. This conference was very square with a lot of Phd's, yet it drew a crowd of around 300 graziers. In 1980, Stan Parsons and Allan Savory's work in the Western United States came to my attention, and in 1981, thanks to electric fence pioneers, Art Snell and Bob Kingsbery, New Zealand dairyman, Vaughan Jones, was our featured speaker at the annual Grazing Conference. This conference drew a crowd of around 600 graziers, but was to mark the end of our traditional "state of the art" orientation. Little did I know that we would not attract another crowd this large for a <u>full ten years</u>.

Vaughan's enthusiasm for building a farm or ranch from the grass up first drew my attention to the possible profitability of intensifying grazing management. In 1982, I traveled to New Zealand where Vaughan served as my tour guide. The experience shattered all of my previously held paradigms about how to raise cattle and sheep.

While coming home on the plane from New Zealand, I overheard an American ask his Kiwi seatmate what he did for a living. "I'm a grass farmer," he said. That simple explanation hit me like a lightning bolt. **This surely was the secret. We were not here to produce meat and milk. Our job was to harvest the crop of grass.**

The New Era

As soon as I arrived home, I changed the name of the magazine to **The Stockman Grass Farmer** and started a crusade to put the grass first in farming. For five years, this was a very lonely course. I was no longer interested in the "state of the art" as it then existed. I was only interested in what "could" exist.

Many of my readers were convinced I had been brainwashed in New Zealand. Others thought I had gone mad. In those early days I could have held our grazing conferences in a phone booth. Advertising withered as we stopped covering the seedstock business. The circulation of the magazine dwindled to a few thousand as our new direction made many former readers very uncomfortable.

In these deflationary, post-Tax Reform days, it is hard to remember how complacent beef cattle farmers in the late 1970's and early 1980's were. They weren't making any money on their cattle, but land inflation allowed them to borrow a comfortable lifestyle with virtually no effort. **Comfortable people do not create change, nor do they appreciate people who are trying to create change.**

By the mid-1980's, **The Stockman Grass Farmer** was bankrupt by every measure except admission. It was able to keep going solely due to the beneficence of our printer who kept extending us credit long after anyone else would have pulled the plug.

Problems Become Opportunities

Ironically, it was land deflation and the 1986 Tax Reform Act, that saved us. By 1988, there was starting to become a sizable crowd willing to listen to a form of farming that made good, taxable income. Prior to that time, I had been forced to rely upon research station results and New Zealand examples for editorial. Also, about this time the real results were starting to dribble in from the field, and they were better and more exciting than I had ever dreamed they would be. It is in this era of the late 1980's, that the stories in this book begin.

These early success stories, which have appeared in **The Stockman Grass Farmer** between 1986 and 1993, were very important for both their educational and inspirational reasons. The learning curve is an inverse curve. Things necessarily get worse before they get better because we don't know what we are doing. **A skill is just like speaking a foreign language or learning to swim. One must do it badly before one can do it well. With grazing, this learning curve is at least three years long.** It is much easier to stay the course during these frustrating early years if there are others who have been down the road before you who can serve as an example and mentor. America is truly unique and truly blessed to have a long tradition of pioneers willing to show the way for those who want to follow.

Thanks to Amtrak's cheap All-America fares, I was able to get around the country and keep the success stories coming. With cash perpetually short, I was probably the only editor in America who slept on the couch of nearly the majority of his readers. Like most things seen in retrospect, these were both the worst of days and the best of days. Needless to say, poverty kept me close to my readers.

Today, with the critical mass finally having been reached and the paradigm rapidly changing. I hope these early stories will serve as a reminder of how far we have come in a few short years for the "old hands", and as a source of inspiration and perseverance for new graziers. As Andre' Voisin said fifty years ago, and these stories prove, grass farming can provide you with a "pleasant life in the country."

Go for it!

Chapter 1

Grassfed Beef in Virginia

Joel and Teresa Salatin live in a beautiful white two-story, 18th century farmhouse hard against the Blue Ridge Mountains in the pastoral Shenandoah Valley near Swoope, Virginia. The whole layout, the house, the kids, the roaring fireplace, the home-made apple butter on the table, all seem lifted out of a 1953 Norman Rockwell painting. Even the idea that a couple could raise a family and enjoy a good middle class life from 95 acres of grassland seemed something from a simpler, dimly-remembered time in the past.

Joel's family moved to Virginia to be near Washington, D.C. His father had sunk his life savings into a thousand-acre farm in Venezuela and it had been confiscated by a left-wing government in 1959. The Virginia location allowed him to keep the pressure on the Venezuelan government for payment of the farm. (A token payment was finally made.)

In 1961, Joel's dad bought the farm that Joel runs today. His dad was ridiculed by the neighboring farmers for saying that he

believed the time would come when he would be able to run a cow to the acre on his farm.

Today, Polyface Farms (farm of many faces) runs 100 cows and yearlings on its 95 acres of grassland. This is done without one drop of commercial fertilizer or soil amendment. The only pasture input since 1961 has been rotational grazing and the careful collection of manure and its return to the fields.

"My dad became interested in rotational grazing in the early 1960's. In the 1950's there had been a lot of research into the benefits of rotational grazing in Virginia and dad had seen those results," Joel explained.

However, his father soon became disenchanted with both the cost and inflexibility of permanent post and wire fencing, and being a consummate tinkerer he invented his own system of temporary electric fencing which Joel still uses today.

The temporary electric fences allowed them to increase the stock density per acre and improve the nutrient cycling and buildup in their pastures. Joel's pastures are a rich salad of fescue, orchardgrass, native warm season grasses, and red and white clovers. "By stockpiling the fescue in the fall and then rationing it out in the winter, hay consumption is cut to between 1000 and

12

1800 lbs. per cow." All hay is fed in a barn that is constantly bedded with fresh sawdust and woodchips to capture the manure and urine from the cattle. This manure pack is cleaned out once a year, composted, and spread on the hay meadow to recycle the nutrients.

No protein or grain is ever fed with the hay. The cattle receive geothermally dried Icelandic kelp and salt as a mineral supplement and are wormed with either diatomaceous earth or Shaklee's Basic H. (Shaklee makes no claims for use of Basic H as a dewormer.)

Such organic techniques provide a unique marketing niche for Joel today, but were used for over 20 years as a least-cost way of producing commercial beef cattle. Joel's grandfather was a charter subscriber to "Organic Gardening" and his favorite uncle had the Midwest's largest free-range layer operation in Ohio. So, you could say organic grass farming came "naturally" to Joel.

His cows are timed to calve with the spring grass and only heifers who have never had lice or grubs are retained for the herd as Joel believes this can be eliminated through genetics.

All paddocks are watered with either fresh water in a trough or a running stream. No pond watering or swimming is allowed for the cattle.

Number One Tool

No paddock is more than 200 yards from a tree to promote the natural insect control of bird predation, but **the cattle are not allowed to "shade up" under the trees as this would break the nutrient cycle and provide breeding grounds for flies and disease.**

Shade is provided by a homemade, portable "shademobile." This mobile provides 1000 square feet of shade and is built around an old mobile home frame. It is supported on a tricycle style mount and all three wheel sets can be made steerable to get through tight spots.

Of all his soil building techniques, Joel credits the "shademobile" as being his number one tool. By parking the mobile on a thin area he can concentrate the herd effect, manure, and urine in a small area. He moves the "shademobile" daily (twice daily in very wet weather) to spread the manure evenly and prevent permanent

pasture damage. A second "shademobile" will allow him to go to a leader-follower system with the yearlings grazing in front of the cows and calves.

Since 1961, the cowherd has been completely liquidated three times and built back with day-old, bottle-fed calves. The current cowherd is predominantly angus and is being bred to a Brangus bull. Until 1988 Joel used A-I, but was not satisfied with the type of selection the A-I studs were using and decided to go to natural service. However, the addition of a bull has complicated his system.

As a result, he is starting a "bull-sitting" service whereby he pastures and cares for his neighbors' bulls in the off-season. "I figure one is as much trouble as a dozen," he said.

Joel continues to feed hay until his pastures have reached a height of six inches in the spring to keep the dry matter up and prevent scouring. He then uses a very fast rotation with a rest period of ten days or less to move over his paddocks. Some of these paddocks are shut up and cut for hay so that the grazing pressure can be kept up on the others and they can be kept young and vegetative.

"I hay more to keep pasture quality than to make winter

feed. I've already got four years' worth of hay in the barn, which is a pretty big gap in my nutrient cycle," he said.

This building hay surplus is one reason for the bull-sitting service and a reason for adding purchased stockers to his class mix. Joel has his grass program tuned to producing a 1000 lbs. slaughter steer at 18 months of age and a 950 lbs. slaughter heifer at 24 months. Steers are finished on the spring pasture and are slaughtered in June and the heifers are slaughtered the following fall.

The slaughter animals are sold at the farm on a liveweight basis for $84 cwt. for steers and heifers and $55 cwt. for cull cows to his customers. He then delivers them to a nearby local slaughter plant for the customer.

After all cash costs and labor are deducted, Joel nets between $600 and $800 per head.

All offal is returned from the slaughter plant and is composted and returned to the soil. He is also considering tanning his own hides in the pioneer way for a local bootmaker. "When you consider that a pair of boots lasted for the entire Civil War tanned the old way, I think we have lost something," he said.

Tanning is a pretty high value added process too, Joel said. A $35 hide sells for some $450 after tanning.

All slaughter cattle are contracted four months before slaughter and he has a two year waiting list for organic grass-fed beef.

Elsewhere On The Farm

However, as much as it pains me to say this, Joel said that selling organic chicken is easier than organic grass-fed beef. "People will gladly drive over 200 miles to buy a fresh organically grown chicken," he said.

In 1988 he raised and sold some 5000 organically grown broilers. In 1989 he planned production to increase to 7000. The broilers are only grown in the warmer months of the year. (See "Pasturing Poultry Pays" for details on Joel's pastured poultry operation.)

In addition to adding organic egg production due to customer demand, Joel is experimenting with grazing rabbits in portable pens like the chickens. His customers are also asking for organically grown lamb and goat but his fences are not yet up to

15

these two species. However, of all the critters on his place, it's his earthworms that he is the proudest.

We walked out in the pasture and he knelt down and parted the fescue clumps so I could see what he was so excited about. The soil was covered with little piles of their castings (manure). "Twelve earthworms per square foot can build six inches of good topsoil in just two years. Voisin said the life that is capable of being supported above the soil surface is in direct proportion to that being supported below the soil surface. Of all the animals I am trying to care for on this farm, these little critters are probably the most important," he said.

(Editor's Update: Joel now slaughters all of his beef animals in the late fall. He believes this fall-slaughtered beef is tastier and, of course, the animals are larger and more profitable.

The grazing of the rabbits in cages has become a profitable enterprise for Joel's son Daniel.)

Chapter 2

Pasturing Poultry Pays

If you are looking for an additional grazing enterprize that can return some big bucks per acre, you may want to investigate pastured poultry.

"A pastured poultry program like ours can be an excellent way for a young couple working six months a year to net $25,000 for their labor," said Joel Salatin. He and his wife, Teresa, operate Polyface Farm near Swoope, Virginia.

Salatin's return to the acre that is actually being grazed is between $1000 and $1500. But keep in mind that **the only grazes his chickens across an acre once every three years.** This long rotation is to prevent any disease pickup by the chickens and to allow the pasture to fully recover.

"Chickens are really hard on a pasture," he warned. His pastures are predominantly fescue and orchardgrass with white and red clover.

Joel said pasturing poultry can cut at least 30% off the chickens' feed cost and two weeks off their grow-out time. Capital

expenses for housing are low and health problems are practically non-existent with the long rotation.

Salatin also raises and sells grassfed organic beef. He said a pastured poultry operation needs to be combined with ruminant livestock to optimize the productivity of the land.

The acre of land grazed over by the chickens in May and June will subsequently have two cuttings of hay made from it and be stockpiled for wintergrazing by the cows and yearlings as well.

One acre of pasture will graze the equivalent of 500 birds on one pass across the pasture. Death losses can be severe in cold wet weather. Salatin said **pasture broilers should only be raised during the warmer months of the year and only on green pasture with legumes.**

In west central Virginia, Joel can raise seven batches of birds per season. Each batch takes eight weeks to grow. Graziers farther south can grow more batches and those farther north fewer to best match their grass and climate.

Joel's chickens are controlled on the pasture by a 10 foot by 12 foot by 25 inch high, wood and aluminum, homebuilt cage. A cage this size will hold a maximum of 100 birds. The pens have a removable feed trough and a commercial gravity-fed waterer that feeds from a small tank on top of each pen.

Pens are moved each morning to fresh pasture. He inserts a small two-wheel dolly under one end and pulls the pen forward slowly from the other end. The chickens walk along with the pen.

"Commercial broiler growers who visit us always comment that there is no smell, no flies, and the chickens are much calmer than their hot-house raised chickens," he said.

It takes a minute and a half to service each pen with feed and water. In the late afternoon he returns to check on each pen's water supply. A large water tank mounted on an old manure spreader serves as a portable reservoir and is taken back to the barnlot every few days for refilling.

The pasture should not be lodged over and should contain a good percentage of clover--the chickens' preferred plant food. **Ideally the chickens' pasture should first be lightly grazed by beef cattle to a residual of four to five inches. The pre-grazing with cattle will both condition the grass and leave manure which is relished by the chickens and is a good protein, vitamin and mineral supplement.** He admitted the coordination of cattle, chickens and rotational grazing was very challenging.

A local feed mill works up Joel's feed ration to his own formula, which includes both kelp and a probiotic. **Joel uses old oil tanks as miniature grain silos and lines them up along the pre-planned grazing path of the chickens. This prevents him from having to carry the feed very far.** The feed company's truck periodically comes out and fills these silos. The empty silos can be easily rolled to the next location.

The manure from the chickens enriches the soil, and their path across the pasture is painted with a bright green regrowth. "Each acre of chickens consumes the equivalent of one acre of 100 bushels of corn. Grain fed on pasture is the same as adding fertilizer," he said.

Pastured Egg Production

Joel finds a major constraint to small poultry producers is access to a hatchery that will agree to service small producers. Large poultry companies have contracted for virtually all the hatchery capacity in many areas.

He sells the birds live to his customers, then custom slaughters them on the farm in a low-tech backyard processing shed. This live sale/custom slaughter sales sequence is necessary due to government health regulations. Birds are grown to 4.6 lbs. live-weight, and are sold on a dressed-weight basis for $1.15/lb.

The newest addition to the Salatin's pastured poultry operation are laying hens. The same customers for his chicken and beef also wanted eggs, so Joel has added them this year with the construction of a portable hen house he calls the Eggmobile.

The Eggmobile is a 10 x 20 foot wooden hen house built on a set of mobile home axles and is designed to accommodate 100 layers. The house has laying boxes built around the edge that can be top-opened for easy removal of the eggs.

The floor of the hen-house is wire mesh to deposit the hens' droppings on the pasture. A small removable ramp allows the chicken to enter and leave the house. The chickens naturally return to the house at dusk and will graze no farther than 20 to 30 yards from the house so no daytime fences or containment is necessary.

The Eggmobile is moved four days behind the beef cattle in their pasture rotation. The four day lag is timed to optimize the hens' eating emerging fly maggots in the cattle's manure.

The amount of insects the hens are able to glean from the manure and pasture has cut the feed consumption of the hens by 60% with no decrease in egg production. The hens are given whole-shelled corn free-choice and have access to some meat scraps for supplemental protein.

"The big expense in egg production is protein not starch. This way we let the hens gather their own protein and also reduce our fly and insect problems," Joel explained. The hens' scratching breaks up the cowpies and encourages them to break down and recycle faster.

Joel said he has had no predator problems with his hens and thinks the frequent moves have something to do with this. "Predators don't attack on the first night. They like to check things out first. I think the fact that the hen house moves every day or so keeps them confused and wary," he said.

Unlike his broiler chickens, Joel plans to run the Eggmobile year-round.

(Update: Joel has written an in-depth book on his methods of raising pastured poultry.)

Chapter 3

Buying the Farm--Fast

Can cattle pay their way? I mean, can they actually buy a farm for you? Harland Rogers, owner of Rogers Bar H R Ranch near Collins, Mississippi, says yes. How long a payout? Would you believe six months? For everything? You bet!

Rogers runs a combination registered Charolais/commercial stocker operation on 1600 acres of grass in the hilly Piney Woods region of south Mississippi. "You know," he said, "we used to be ashamed to go into the bank with our cowboy boots on, but I think we can wear 'em in there with pride now."

Rogers had been intrigued with the idea of controlled grazing since he heard a speech by New Zealander, Vaughan Jones, at the Southeastern Grazing Conference in 1983. He kept reading the success stories in the **Stockman Grass Farmer** magazine and finally decided he had to either prove or disprove these stories to himself, once and for all.

As a farmer/director of MFC, a regional farmer co-op, Rogers proposed a test of controlled grazing on a recently purchased 80-acre tract of cutover timberland near his ranch. All cattle were to be weighed on and off the tract and scrupulous records would be kept on all expenses. Rogers, who uses broiler litter as fertilizer on his own ranch, agreed to fertilize to make the comparison more typical to other stocker operations in the area.

The 80 acres were plowed and seeded to a pasture mix of oats, Marshall ryegrass and vetch. An excellent stand of Crimson and Arrowleaf clover volunteered from a previous planting. A ten-paddock cell was built in a partial radial design with fiberglass poles and electric polytape.

Building Stocking Rates

An initial stocking of 139 head of four-weight steers on December 1 was quickly determined to be too light. An additional 42 steers were added on December 12. This was also determined to be too light and 22 Charolais bulls were added. The bulls were moved off on January 15 and 14 more steers were added on January 25. 24 more were added on March 17.

Rogers bought number 2 and 3 Brahman-cross steers that averaged $56 cwt. and contracted them for late May delivery at $63 cwt. Rogers likes the Brahman-crosses for their health and growthability. "I've found that buying a plainer sort of calf and upgrading him on ryegrass is a lot more profitable than buying a premium-priced number one black-baldy," Rogers explained.

Rogers said that by contracting early he left about four dollars on the table but said he would always take a profit when it was offered, rather than take a chance on a wreck like the 1986 dairy buyout collapse.

The steers produced a gain of 684 lbs. per acre for a gross sale of $431.06. Interest, fence, hay, land prep, seed, shots and implants came to a total of $151.04. The forward margin on the steers' purchase price added another $81 an acre for a total gross margin of $361.64 an acre.

The land cost $200 an acre and an additional $90 was spent destumping and preparing the land for pasture seeding. Subtracting this $290 an acre leaves $72 per acre for management, plus a fenced and watered farm, all from the first six months of production. The cell was to be used for summer crabgrass grazing with heifers following the ryegrass. While these figures are certainly impressive, Rogers warns that they are enhanced by the forward margin on the purchase weight. "I think a man had better figure a negative margin of at least 4.5 cents for a normal year. Using that figure would have lowered our profit per acre to only $175 an acre, which is still not a bad return on $200 land."

In defense of his figures, Rogers said that most of the hay charged against the cattle was used not to feed the cattle, but to plug erosion blowouts on the steep land. Also, the late turnout cost him three weeks of grazing on the front end.

"The saying that intensive grazing is like piloting a jet is certainly true," he said. "You make a mistake and you've gone over three counties before your realize it. We learned a lot this year that should enable us to do even better in years to come."

Rogers said that **graziers using prepared seedbed grasses need to have a small trap of range or sodgrass to move the cattle to during periods of heavy rainfall**.

Learning The Hard Way

"We didn't do this and left our steers on the grass with it raining five inches a day," he said. "We came out okay, but I wouldn't do it that way again. Also, I would take a couple of portable reels and force my cattle to graze the bottoms first and save the hillsides for wet weather."

Rogers feels the use of temporary fencing is also a good idea until you get a feel for how your cell should be constructed. The cell should ideally be designed to allow the separation of wet and dry land and the paddocks should be as square as possible, rather than long and narrow, to prevent erosion and trailing damage. Stocker cattle should also be broken to electric fence before being introduced to a minimally fenced cell.

Rogers' son, Bernie, who ramrods the 1800 head of cattle on the ranch, said that the cattle's intake could be increased by shifting the cattle after their normal morning grazing period.

"There may be areas where intensive grazing won't work, but in these steep clay hills this is the answer," Rogers said. "I've got the figures to prove it."

Costs and Returns
From 10-Paddock, 80-Acre, Grazing Cell,
Roger Bar HR Ranch, Collins, Mississippi
December 1, 1986 to May 15, 1987
All figures shown are on a per acre basis.

Initial stocking rate		955 lbs.
2.2 steers x 434 lbs.		
(later increased to 2.7 steers)		
Gain		684 lbs.
GROSS SALES INCOME		$431.06
(684 lbs. x $0.63)		
Production Costs:		
Hay	$ 20.00	
Electric Fence	8.33	
Grass	81.62	
(seed, fertilizer, prep)		
Animal Health	7.96	
Interest (10%)	33.13	
TOTAL PRODUCTION COSTS		$151.04
Production Margin		$280.02
Marketing Margin--positive		$ 81.62
GROSS MARGIN		$361.64
Cost of Land	$200.00	
Land Preparation	90.00	$290.00
(stumping, water, corrals)		
NET RETURN TO LABOR		$ 71.64
Estimated daily labor for entire cell		30 minutes

Chapter 4

Grass Farm Looks Good
As Potential Retirement Job

Hub Spencer admits that when he first heard controlled grazing he thought it was all a bunch of "hooey." The company he works for, Breeders Supply of Lexington, Kentucky, had just taken on a line of New Zealand energizers and fence components, and Hub thought the claims that were made were just a little too fantastic. Hub and his son, John, first subdivided their old "home-place" farm as much to disprove, as to prove, the fencing company's claims.

"John and I figured if we had to sell these products, we might as well try it ourselves," Hub said.

Hub decided that he would rather run steers in the summer than to continue to hay cows through the north central Kentucky winters. He enlisted in the University of Kentucky's "Graze-More-Beef" project. "Graze-More-Beef" is a UK extension program to help develop a better balanced market in Kentucky by teaching stocker grazing techniques to the state's primarily cow-calf oriented grass farmers. Ken Evans, extension agronomist, and Curtis Absher, extension animal scientist, serve as coordinators for their farmer cooperators.

The Spencer farm consists of approximately 120 acres of open grassland and forty acres of trees. The predominant grass is "dirty" fescue. Learning the tricks of taking cattle through the summer on predominately cool season pastures has come painfully.

First Year Hardest One

"Our first year, we stocked up with 150 steers in the early spring and they were galloping along at over two pounds a day," he recalled. "Near the end of June, Curtis told us to sell half of our steers, but we didn't do it, because we had a reserve of grass we hadn't grazed back in the woods. When our grass got short, I turned 'em in on the rank fescue and the steers fell absolutely to pieces. I learned that year to listen to Curtis."

Absher said that **grazing fescue through the summer was not "put and take," but "put-take-take-take-take."**

"Once the weather gets hot, about the only gain the cattle are making is what they can get out of the clovers. Once the clover goes, you might as well sell the cattle and buy back in when the weather turns cool in the fall," Absher said.

Despite a disappointing first year, Hub and his son decided to put even more money and effort into the farm. They subdivided the farm into 33 paddocks and seeded some of the paddocks to ladino and red clovers.

Following Absher's advice, they decided to go to a "split-turn" with a spring set and a fall set of cattle. In late April, they bought 205 steers and grazed them until the end of June and then sold them. These steers turned in an average daily gain of 1.85 on "dirty" fescue. Another set of steers were purchased in August and were grazed until the end of November. These fall steers had an average daily gain of 2 lbs. a day.

"Most people totally overlook a fall turn of stockers, but we've found our animal performance on infected fescue is higher in the fall than in the spring due to the high sugar content of the grass," Absher said.

Due to excellent animal performance and good buy-sell margins, the second year in controlled grazing has been very

27

profitable for the Spencers. They now plan to expand their cell to 40 paddocks run as two 20 paddock cells.

"We've found that 20 paddocks is our most optimum subdivision," explained son, John Spencer. "Once we got over 20 paddocks we couldn't move fast enough to keep the grass quality up."

In 1987, the Spencers planned to run 300 steers (150 per cell) per turn. They have been pleased with the red clover and plan to add more. Since starting controlled grazing, their paddocks have volunteered excellent stands of bluegrass and native white clover that help dilute the fungus in the fescue.

"It's been amazing to me the way cattle can clean up the land. I never thought I'd say it, but I want a small cowherd again that I can use just to clear up rough land with. I'm also thinking about getting a small herd of goats for thistle and ironweed control," Hub said.

"This is turning into an excellent retirement job for Dad," John said. "He'll be out here opening a gate once a day and making more money than me putting in a full day."

Chapter 5

Balanced Forage Flow
California Grass Farmer's Goal

John and Virginia O'Connell of Flournoy, California, have the same problems of most grass farmers in cool-season grass country. The fall growth comes on too slowly, barely grows at all in the winter, and then explodes into a spring lush that is largely wasted. What John wants (and what we all want), is a more even distribution of forage throughout the grazing season. With this goal in mind, he has been experimenting with annual leys (temporary pasture) and baleage. The O'Connells run 1000 to 1500 steers on a gain contract each winter and spring and also grow around a 1000 acres of dryland wheat.

"I primarily have three goals," John explained. "I want to be able to bring in stocker cattle earlier to help my clients on their purchase price. I want to provide a better forage quantity during the "crunch season" in mid-winter, and I want to better utilize the spring lush."

The native cool-season annuals provide too little fall and winter growth so John has been supplementing them with annual ryegrass and crimson clover no-tilled into his native range.

"We have been successful growing sub-clover here, but its growth pattern only adds to the spring lush and therefore doesn't solve our forage distribution problems, like the early growth of Crimson," John said. **"You've got to put your money in plants that grow out-of-season or you might as well stay with native range."**

John is also experimenting with Kennel Festolium--a cross between fescue and ryegrass--and with Ellett perennial ryegrass from New Zealand.

"We are tickled with the prospects of Ellett. We can get higher dry matter production per acre **and** better palatability than with our annual tetraploid ryegrass," he stated.

John is deeply impressed by the work of Newman Turner, an English dairyman, who published a book in 1955 called **Fertility Pastures**. In this book, **Turner recommends that annual leys be designed to enhance the livestock performance and improve the soil's fertility at the same time.** These "pasture salads" can contain as many as 15 different species of grasses and legumes. These leys are used as supplements to perennial pasture during the growing season to keep milk production up and are cut as silage for the sole mid-winter feed source for his dairy animals. Newman believed these leys not only balance nutrition but also bring the soil into balance as each plant concentrates a different mineral element in the soil.

Planting Annual Leys

"We've been experimenting with various ley mixtures and have some 30 to 40 test leys scattered around the farm to see how they will grow on our various soils," he said. A typical ley might contain Birdsfoot Trefoil, Plantain, Yellow Blossom Sweet Clover, Austrian Winter Peas, Alaskan Peas, Loma Vetch, and several varieties of Medics.

"What we are trying to build is that full season temporary pasture that we can both strip-graze with stockers during the growing season and preserve as baleage for use in the later summer and early fall, when our pastures are still dormant."

John was an early convert to controlled grazing, but feels that the typical 10 to 12 pasture permanent cell is not intensive

30

enough for stocker cattle on planted annual leys. He is rapidly switching to moveable fences and strip-grazing.

"A cell that has enough division for the mid-winter rationing period has too many for the rapid spring growth period and vice-versa. **By using temporary moveable fences, you can more accurately match the animal pressure to the growth of the grass. With baleage, we can save that high-quality spring lush for use when we need it next fall.**

"What I want is a system that will allow me to guarantee a 250 lbs. per head gain over a six month period to our clients," John said. "We aren't there yet, but I think we are getting close with our current approach of concentrating our attention on providing pastures that can put a pound and a half gain on cattle for the full season. I have never been more excited about the future of agriculture than I am today. I don't want any of the doom and gloom guys around me."

John has been collecting extension publications from the 1930's through the 1960's and has found them an excellent resource material for today. He also recommends Louis Bromfield's books from the 1940's on the development of "Malabar Farm," a state-of-the-art grass farm in the post World War II period.

John concludes, "The research has been done. The technology is here. We've just got to go back and refind our way to lower cost production and we can do it."

(Update: Perennial ryegrasses have not been successful in California except under irrigation. Also, the highly variable winter rainfall in California has prevented John from achieving his goal of 250 lbs. per head in 180 days.)

Chapter 6

Oil and Cattle No Longer Mix
On South Texas' King Ranch

King Ranch's South Texas neighbors once described the sprawling 825,000-acre ranch as "the ultimate cowboy playground." Riding, roping and dragging calves to the branding fire in true 1870 tradition continued on the King Ranch long after most ranches had adopted squeeze chutes. The neighbors wondered what King Ranch would be like without the income from their 2,730 working oil and gas wells. Today they know.

In 1977, the ranch's oil and gas revenues were diverted from the ranch account and paid directly to King Ranch heirs. The ranch management was told to make the ranch self-sustaining and profitable from its cattle and farming operations. This shift brought an end to some of the cowboy traditions and is still being reflected in management changes.

One King Ranch hand described turning the ranch around as being similar to turning an oil supertanker. "It takes a lot of water to turn one of those big babies and the King Ranch can't turn on a dime either," he said.

Since 1977, King Ranch has been dissolving its vast overseas holdings. Its Mississippi ranch has been sold; its Florida ranch plowed up and put exclusively to sugar cane. On King Ranch some 37,400 acres of good blackland has been put to cotton and milo, and the 15,000-head capacity Estaban feedyard was opened to outside investor feeders on a custom basis.

Quarter horse breeding was centralized at one division and turned into a profit center. The cattle division has been separated into purebred, commercial cow-calf, and stocker-feeder areas, with each having to pull its own weight. The ranch's feedlot has to buy its milo from the farming division. The stocker-feeder division has to purchase calves from the cow-calf division and pay interest on those purchases. On the Norias division, corporate hunting leases have become a bigger source of income than cattle.

King Ranch has never used barbed wire. Its famous 2000-mile net wire fence is supplemented by New Zealand power fence. Three grazing cells are in operation. Working pens are being consolidated in fewer locations with lanes leading to them, allowing an easy trail drive to the pens. Helicopters are replacing horses for roundups. Fire is replacing root-plowing and chaining for brush control. The "cowboy playground" has become a business.

"There were a lot of things you just couldn't do anymore when you put the pencil to it," explained Scott Kleberg, assistant manager of the cattle marketing division.

The New Generation

There were some things that the ranch **had** to do with the narrow margins in cattle. For example, all 50,000 cows are pregnancy tested every year and the open cows sent to slaughter. All bulls are semen tested every year and the sterile bulls sent to slaughter. Heifers are beginning to be bred as yearlings to calve as 2-year-olds.

Tio and Scott Kleberg are the "new generation" of King Ranch and both of them are in their 30's. Scott ran his own stocker operation before recently signing on with the family ranch. In King Ranch tradition, Scott and Tio work in the roundups with the Kinenos, as their hands are called. English is used in the office. Spanish is used on the range.

James H. Clement is the president and chief executive office and the first "outsider" to hold that position. It has been under his leadership that the oil and gas royalties have been divided from the ranch income and the ranch forced to sink or swim as a paying operation. Clement's election as CEO by the family stockholders over the two "heir apparents," B. Johnson and Bobby Shelton, set off a mini-Civil War within the corporation. Both have subsequently sold out and started their own ranches elsewhere in Texas.

Holding King Ranch together over the years has been a continuing struggle. Many employees feared a loss of interest in the ranch by the new generation of heirs. Tio's and Scott's strong interest in the ranch's operation has relieved a lot of these fears. The shift in direction to a self-sustaining operation, before the oil and gas wells run dry (in the 1990's), is now seen as a good move, although many heirs openly wondered if the ranch could pull it off. "All of us work under the eye of management's computers," explained Scott Kleberg. "If our division isn't making money, management calls us in and wants to know why and what can be done about it," he said.

Bull Grazing

George Durham, manager of the commercial cattle marketing division, oversees the grazing and feeding of the ranch's progeny. All King Ranch's male progeny are grazed and fed as intact bulls. This allows bulls, which show exceptional gains and conversion, to be sorted out and used as seedstock bulls. Bulls also show better gains and conversion, Durham said.

The bull feeding started as a contract operation to supply lean beef for McDonald's hamburgers. McDonald's was test marketing a new steak sandwich in the Southeast. Unfortunately, the sandwich didn't fly, but the King Ranch supplied some 200,000 bulls during the tenure of the contract. Today, the bulls are slaughtered in Corpus Christi and marketed on the East and West Coast, where supermarket chains have their own "lean beef" programs. These bulls sell for some $2 to $4 under Amarillo steers, Durham said.

Most of the bulls go on feed from 600 to 650 pounds and are fed for 140 days on popped and rolled milo. Lightweight calves are sent to the Marlin, Texas, area and grazed on oats on a cost-of-gain contract. These yearlings usually go on feed in June at around 800 pounds to help keep the year-round capacity of the feedlot up.

A new little "profit center" for the feeding operation is a manure composting plant. Feedlot manure is composted and then bagged and sold as "King Ranch Cow Manure" through yard and garden shops and supermarkets. "We now get $50 a ton for manure (wholesale) that we used to beg people to haul away," Durham said.

Most of the grass on King Ranch is their own bluestem, which was genetically adapted from a South African grass. No hay is put up and the cows overwinter on "standing hay" and range-cubes from their own feedmill.

Purebred Operation

Scott Kleberg said most of the cows on King Ranch weigh about 1,000 to 1,100 pounds, which he considers "optimum" for a rangeland operation. While most of the cows are Santa Gertrudis, the ranch does have some Brangus and Braford commercial cows. "We found it useful to the purebreds division to be able to take a rancher out and show him what a Santa Gertrudis bull on his type of cows could produce," Scott said.

Approximately 2000 cows are in the seedstock unit. From these 2000 some 500 head are picked to be a part of the ranch's "A herd" from which foundation Santa Gertrudis are sold.

"With a genetic base of some 50,000 cows we are constantly able to find cows with exceptional genetics that can move up into the seedstock operation," he said. Cattle in the seedstock division are restricted to no more than a 60-day calving season and must breed as yearlings and then breed back as 2-year-olds.

"That's what the cattle business is coming to," Scott said.

He also said that the ranch had tried Coastal bermuda, but found it unsuited for a rangeland operation because of the high fertilizer requirements. The bluestems are good cow grasses. However, they are limiting to a post-weaning program with a July-August weaning season. "We can grow oats down here, but the deer ate more of them than the calves. We found deer don't like wheat, but our heat prohibits wheat planting until December. I

35

guess you could say we're still searching for a good post-weaning grass or legume that will fit in our South Texas climate," Scott said.

Scott stated that they have been pleased with their grazing cells, having lived up to their advertising for raising the stocking rate, performance and grass composition. "We are using the cells right now for growing out replacement heifers. I guess you could say we are still kind of boggled with the idea of cross fencing 825,000 acres."

Fire is becoming a major management tool on the ranch for mesquite control. "We would like to go into a planned 3-year burning rotation on all of our land, but sometimes drought gets in the way. We have found that fire will not eliminate mesquite but it knocks it back to where it is not a major problem," Scott said.

With no hay, silage, or backup roughage source and a highly variable rainfall, the ranch uses a conservative stocking rate to allow it to get through drought years. "Last year we were burning prickly pears for the cows and this year they are splashing around in water 6 inches deep. That's the Texas Gulf Coast for you," Scott said.

In fact, when I visited King Ranch in late May 1985, it looked more like South Louisiana than South Texas, a fact for which the ranch hands seemed almost apologetic. "You should have been here last year and seen what cattle ranching was all about," they said.

What's happening on King Ranch today is probably symbolic of the whole cattle business. We're all growing up and becoming mature. We're proud of our cowboy past, but have to answer to modern computers and conditions. We've got one foot in 1870 and one foot in 2001.

King Ranch is into embryo transfer and A.I. on the Santa Gertrudis division, but the southern most Norias division is preserved as a rough and tumble, ride 'em and rope 'em and drag 'em to the branding fire reminder to current and future heirs of where they came from and what their forebears had to endure.

Current management makes no apologies for the continuation of this tradition at Norias. In the fact sheet given to visiting journalists they write:

"The skilled vaqueros still work the mesquite-choked brush country of South Texas just as they did a hundred years ago. Many things change, the methods, the technology, the systems of management and even the look and productivity of the cattle, but at this level, it is still man and horse against the country, and in the foreseeable future it will be."

Chapter 7

Relearning How to Farm

Jim Whitfield moves the forward polywire in his wheat, ryegrass and clover pasture and the sows surge forward to gobble down the fresh grass and legumes. He then moves the back wire to prevent regrazing until the pasture is ready to be taken down again. This strip grazing process continues 12 months a year on small grains and ryegrass, and on crabgrass in the summer. One electric wire provides the only permanent fencing for the sows.

"This is nothing new," Whitfield said. "This is the way we used to raise hogs in the South. We've just forgotten how to farm in the last 20 years. The primary difference between the way we raise feeder pigs and everyone else does it is about $20 a pig. I had a farmer tell me he'd go broke before he went back to pasturing pigs. I think he'll definitely get his chance. You've got to let a hog be a hog. There's no nutrition a sow can get from a concrete floor."

Whitfield "partners" on the sows with John Hines, a diversified farmer, in the Delta part of Yazoo County, Mississippi. The Hines plantation also has a purchased feeder pig finishing

operation. "The universities will tell you that a pasture pig operation is a low cost operation, but produces a small litter size," Whitfield said. "Our average litter is 10 to 12 pigs. The sows breed better on pasture and are much healthier than confinement hogs."

The pasture mix consists of wheat, ryegrass, crimson, ladino, and wild winter peas. "The wild winter peas are the sows' favorite. They'll eat all the peas out of a paddock before eating anything else," Whitfield commented.

Sows are farrowed on pasture with the exception of the mid-winter months and will eat approximately a ton of feed each year in addition to their pasture. All the corn for both the feeder pigs and finishing operation is grown on the farm. Quite a few of the finished hogs are sold through the plantation's store retail, and a small cattle finishing operation has recently added beef to the store's meat mix. **"We added $90 a head profit to our hogs by selling them retail through the store. This is something I believe every small livestock producer should look at," said Whitfield.**

The hogs and beef cattle are slaughtered at a state-approved slaughter plant. Vegetables grown on the plantation are also retailed through the store with a pick-your-own turnip and mustard greens patch proving very successful in winter. The mustard greens are also fed to unthrifty pigs in the finishing operation to improve their health. Whitfield said they have seen a tremendous benefit in the health of their purchased feeder pigs from feeding supplemental fresh mustard greens.

The turnip patch is another experiment Jim is watching closely. In the fall 10 tons of composted gin trash was plowed into the turnip patch to see what the combination of a high level of organic matter and the naturally rich Delta soil would produce. As of mid-January these greens were colorful, leafy, and tender with no frost or insect damage. The greens grown next to the high organic patch, however, were dead.

He reached into the patch and brought up a handful of dirt. Despite the heavy winter rainfall, the dirt crumbled in his hand. "You can come out here in this turnip patch and walk across it after a rain, whereas, you'd bog to your knees in a cotton patch. **We've found that standing water in a field is a prime indication of low organic matter.** Last year we put out 300 tons of composted gin trash and imported over 700 tons of chicken manure for our use and our neighbor's use. I'm convinced that low organic matter is the primary yield limiter in the Delta."

Whitfield attributes this low organic matter to the Delta's tradition of heavy use of anhydrous ammonia and continuous cropping. "Anhydrous ammonia sterilizes the soil and kills all the earthworms and other beneficials in the soil. The amount of insects above the soil is indirectly proportional to the life below it. **Each year we followed our anhydrous application with chicken manure to re-inoculate the soil with life and had no insect problems at all in our corn,"** Whitfield stated.

As for continuous cropping, Whitfield shakes his head. "The tradition here is to only work six months a year. Everything is shut down and everyone is laid off for the winter. We're virtually the only farm in our area with some kind of year round operation. The "set aside" program is ideal for someone wanting to go to winter-grazing, but that would mean winter work and people here just aren't used to that. We're gearing our operation for the low prices we believe are coming, not the high prices of today," he said. "A pasture system is definitely the low-cost way to go."

(Read on for details of Whitfield's low-input cattle operation.)

Chapter 8

Cows, Clover, Compost
Revive Cotton Country

Cattle in the Delta region of Mississippi are now so rare that many motorists slow down to take a closer look at the 200 heifers wintergrazing on Jim Whitfield's clover-covered, cotton fields near the Yazoo River.

Only a single wire temporary electric fence separates the heifers from the highway and a circle of cotton trailers serve as a temporary corral.

"I think we would definitely qualify as a low-input cattle operation," Whitfield said.

Whitfield and John Hines farm around 1500 acres of cotton, corn, wheat and soybeans and a small pig finishing operation. Ignoring the taunts and disapproval of their neighbors they have decided to march to a different drum and rebuilt their farm from the soil up.

Out in the broad flat fields, a manure spreader works spreading rich, black, cotton-gin waste compost. Like the cows, the manure spreader and the compost is an oddity in the Delta where

farmers prefer to buy their fertility in an anhydrous tank.

Whitfield's low-input cattle operation grew out of their difficulty in plowing down their cotton land cover crops each spring. "We had Crimson clover nearly waist deep each spring. It would grow and fall over and grow some more. I had some clover plants that you could pick up and they would be six feet tall," he said.

Whitfield had been grazing sows on the clover and vetch cover crops but they barely made a dent in the rank forage.

In the autumn of 1988, Whitfield put the word out that he was interested in grazing stocker cattle on the gain. A group of Louisiana investors approached him and asked if he would be interested in grazing thin slaughter bulls for them.

A deal was struck at 30 cents a pound and during the winter of 88-89, Whitfield put 300 lbs. of gain on 150 slaughter bulls in just 100 days by stripgrazing his cover crops of Crimson clover, vetch and cereal rye.

"We grossed $90 a head in 100 days from a crop that we had formerly just plowed under," he said.

In 1990 the same group of investors came back and asked him to graze stocker and replacement heifers for them.

The Delta region of Mississippi is a barely drained alluvial swamp and farmers have backed away from grazing winter annuals due to fear of extreme bogging problems. In 1990 the Yazoo City area received over 30 inches of rain in just three months and Whitfield got a good lesson in dealing with cattle bogging.

"A lot of people are afraid of intensive grazing in boggy soils and think it will make the situation worse. It was our experience that while we did often bog a field unmercifully going across it with our strip-grazing, it had almost fully recovered within three to four weeks.

"One of our landlords asked why we hadn't grazed one of his fields and was surprised when I told him we'd grazed it just three weeks earlier. He couldn't believe it."

His stripgrazing technique is to move the front fence every day or so and the back fence every five or six days. All fences are polywire on solid round fiberglass posts.

With the exception of three or four days during the severe December freeze that hit the Eastern half of the United States in 1989, he had fed no hay at all.

Whitfield was able to do this despite the fact that the farm has no permanent pasture and no permanent fences at all. He uses no nitrogen on the cover crops and charges only the seed and seeding cost to the cover crop. He believes a lighter than normal stocking rate of one yearling to every two acres is best because of the exceptional average daily gain it can produce.

"We have to pull the cattle off in early April so that we can get our cotton in and this prevents us from using compensatory gain to the extent other graziers do. We have found that the key to keeping your investor happy is a large gain per head."

Wylie Johnson, one of Whitfield's landlords, said that he whole-heartedly approved of the cover crops and grazing on his land. "I've noticed when it rains, the water running off these clover covered fields runs clean and clear while everyone else's runs muddy."

In the winter of 1990 Whitfield spread some 1500 tons of cotton gin trash compost at a rate of two tons to the acre. Cotton gin trash are parts of leaves, burrs and some lint left after the cotton is ginned.

Each ton of gin trash contains 34 lbs. of N, 25 lbs. of P, 35 lbs. of K and 50 lbs. of Calcium and is available free at the gin. Whitfield adds pig manure slurry to his gin trash, but that gin trash has enough natural nitrogen in it to compost without adding manure. By turning it each week, the gin trash totally composts in six weeks. The less frequently it is turned, the longer it will take to compost. When it is fully composted, the gin trash is a deep black with an appearance similar to soil.

After composting he figures it is worth $40 a ton as fertilizer and he will have $15 a ton in it by the time it is spread. Whitfield is as interested in the organic matter building properties of the compost as the fertilizer analysis.

Due to environmental restrictions, gins are not going to be able to burn the trash and each year Mississippi generates some 150,000 tons of gin waste.

Lime as Fertilizer

Another old trick Whitfield has discovered is the fertilizer value of calcitic lime. The Delta's soils are generally high in pH and therefore seldom, if ever, limed. Whitfield read research done at the University of Missouri in the late 1940's by Dr. William Albrecht that showed the fertilizer value of lime was more important than the pH enhancement. After reading this he decided to try liming his fields.

In 1989 when soybean yields for most of his neighbors averaged only 20 to 25 bushels, Whitfield's limed, composted and grazed fields produced the highest soybean yields he had ever grown--40 bushels per acre.

"From the response we got from the lime, it appears that Albrecht was right. Our pH was high, but we were still Calcium deficient. I think we should all start thinking of lime more as a fertilizer than as a pH changer," he said.

After a visit with Jim, you realize that there is a lot more going on in his cotton fields than one can see from the road.

Chapter 9

Drought Proves Benefit
Of Perennial Pasture System

The severe drought that whammied the nation's wheat belt in the fall of 1989 reinforced Walt Davis's belief that annuals are too "iffy" a forage source to be exclusively relied upon in southern Oklahoma.

"It was our experience that wheat pasture provided good winter grazing only two years out of five," he said. "When it was good it was very, very good. But when it was bad it was disaster."

In the last several years Walt has shifted his forage program almost exclusively to perennial forage plants with various companion legume components. Davis now counts some 30 different legume species in his pastures in Bennington due to extreme differences in soil structure and pH on various parts of the farm. Arrowleaf and Crimson clovers predominate on the sandy soils, white clover on the intermediate and Bigbee Berseem in the heavy clay bottoms.

Walt's constant experimentation with various legumes is described good-naturedly by his wife, Diane, as "Our annual financial sacrifice to the clover gods." His primary stocker forage is a mix of alfalfa, fescue, and Johnsongrass and his cow pastures are a mixture of fescue, common bermuda and clover. Ironically, none of the above combinations were planted but were nature's response to intensive grazing. For example, the alfalfa/Johnsongrass/fescue mix naturally came about when he started grazing old thin alfalfa hay fields several years ago.

"I've got a 450 acre 20 paddock, cow cell that was all bermudagrass with the exception of a five acre patch of fescue around a slough. Today, just as a result of grazing management the pasture has developed a percentage of fescue in with the bermudagrass over the entire cell."

This small amount of green winter feed in his cow pastures has allowed Davis to discontinue routine feeding of supplementary protein to his late March and April calving cows. "With our current program we are able to carry a spring calving cow year-round on slightly less than two acres with no feed, no fertilizer and around 1500 lbs. of baled pasture clippings," he said.

Profit In Optimizing Production

Due to the excellent economics of this cow program, Walt has discontinued his purchased stocker program and has expanded his cowherd to over 700 cows. "We have learned that the profit in the cattle business is in optimizing production, not maximizing it," he said.

His cows wean at around 450 lbs. but since they will always be stockered under his ownership, he doesn't care what their weaning weights are as long as he can produce them cheaply enough. "It is our intention to use these high price times to get everything in place so that when the prices inevitably break we can continue to produce a calf cheaper than we can buy it." Walt admitted that he had little stomach for the price rollercoaster of purchased stockers.

His emphasis on "optimization" has extended into his stocker forage program as well. All of the farm's calves are routinely grown out on grass and sold as heavy feeders. Steer yearlings are sold at 800 to 900 lbs., and heifers at 700 to 750.

"We have learned that you can't afford to put fat on in the

46

wintertime. We currently wean our calves onto the alfalfa in the early fall. They get 60 days of excellent grazing on the alfalfa before frost. We then winter them at a half to three-quarters of a pound of gain on fescue and alfalfa hay until spring when the gain comes cheap and fast." Another change has been the time of year he sells his yearlings. Like many graziers in the South, he once sold in late May but now summer grazes his yearlings and sells them at heavier weights in September and October.

"We have found our summer gains are the cheapest to produce. Everyone said they would not gain in the South because of the heat and humidity, but heat and humidity aren't problems. The problem is not maintaining forage quality. With our alfalfa/Johnsongrass/fescue mix we can have excellent quality forage from early spring to early winter."

Walt's direct forage cost per pound of stocker gain in 1990 averaged less than two cents a pound if one charges the original alfalfa establishment costs to the former hay crops.

He has reduced the Brahman percentage in his calves to one-eighth Brahman with little or no detrimental effect on summer stocker gains.

Changes Since The 1960's

Davis Farms was originally established by his father as a straight cattle operation in the 1960's, but the emphasis was gradually shifted to crops and wheat field stocker cattle and the farm's sizeable cow herd shrunk to just 200 head. "Everything we made on the cattle each year we gave back on the crops," he recalled.

Walt eventually left the farm and went into the supply business. In the early 1980's, his family asked him to come back to manage the farm. He agree on the condition that he be allowed to make it exclusively a grassland farm and they agreed.

With neighbors going broke all around them during the farm crisis, Walt decided to grow from his own cash flow and retained heifers and avoid production debt.

"Like a lot of people who have switched from crops to cattle, and particularly those utilizing intensive grazing, we have had to grow back into our acreage. We leased most of our farm to others to crop farm until we could get our cow numbers up to the point where we could utilize the added acres."

This gradual conversion from cropland to grass has given Walt an up close look at the process of plant succession and an appreciation that it can't be rushed. For example, Walt would like to put many of these converted acres into native grasses but has found that low organic matter, farmed-out cropland is not conducive to the growth of such high succession grasses.

"Bermudagrass is an okay grass if it is kept young, but it has such a fast life cycle that it is a use-it-or-lose-it (as far as quality is concerned) proposition."

During bermuda's optimum growing season Walt has to shift paddocks twice a day to keep up with the bermudagrass. However, the growth rate is so variable due to the weather that about the only way to consistently match feed supply and demand is with a haying program or a put-and-take grazing system.

The age-old problem with a put-and-take system is where do you "put" the cattle when you "take" them?

"It appears to me that separate paddocks of the slow-growing, slow-maturing native species would be the ideal 'put-place' to use in combination with intensively grazed bermudagrass," he said. "However, this is still a theory because we haven't been able to get the natives to grow in our farmed-out soils."

He has found this need for undisturbed, high organic matter soils to also be true with the fungus-free varieties of fescue. "When they took the fungus out of the fescue, they unintentionally changed fescue to a high-succession grass. I've lost three crops of fungus-free fescue in a row to summer drought trying to get it established on old, low-organic matter cropland."

Walt has never had any symptoms of fescue toxicity in his infected fescue because it is all grown in combination with a warm season grass and legume. His planting of the pure stands of fungus-free were in hopes he could "stockpile" a greater volume of green winter feed in the fall.

"The primary problem I have with all my legume, perennial program is the lack of cool season grass volume to carry forward into the winter. It may be that I will have to use some fall nitrogen to build a cool-season feedbank. I would hate to break my record of no nitrogen use in nine years, but there may be n

At the time we spoke, he said this cool season ⌐
no big problem because considerable hay was produced as a ⌐⌐
product of keeping the fast-growing pastures young and tender. However, as the farm becomes more fully utilized, hay production will decline and this lack of cool-season volume could become a serious production bottleneck.

Walt has all hay cut from dropped pasture in the rotation. Most of the farm's hay is made in late May and early June when weather conditions are most conducive to hay making. This late spring hay making allows the hay feeding to help in spreading legume and cool season grass seed. At that time the cool season grasses and legumes have already gone to seed, but the warm-season component is at its seasonal peak in quality.

At Davis Farms all big round bales are unrolled before feeding and hay is never fed on top of hay. This practice helps prevent parasitism and promotes better nutrient cycling and seed spread.

"I don't know if we will ever get completely away from hay making in the South due to the need to use haying as a way to maintain pasture quality. However, we have a long-term management goal here of being able to safely cut our annual hay reserve to less than 1000 lbs. per cow."

(Update: Walt Davis still has had no success with the endophyte-free varities of fescue.)

Chapter 10

Upper Midwest
Returning to Its Roots

Dairymen all across the Upper Midwest are turning their backs on confinement dairying and going back to what originally made this part of the country America's dairyland--grass.

The heavy metal salesmen and their university flacks have tried to put the spin on the shift back to grass as a movement of small, marginal producers. Of course, this ignores the fact that Charles Opitz, Wisconsin's largest dairyman, is a leading proponent of the back-to-grass movement. Opitz currently grazes his 1300 lactating cows on just 550 acres of grass. He became interested in grass dairying several years ago when he realized he could no longer pay his feed bill and still meet the notes on his farm near Mineral Point.

"Economic research at the University of Wisconsin has shown that increasing the size and scale of a traditional confinement dairy does not markedly lower the capital investment per cow. There is very little economic advantage in being a large confinement dairy," University of Wisconsin extension agent, John

Cockrell said. "The average cost of production is $13 cwt. in the upper Midwest. On average, all traditional dairymen are in economic trouble regardless of size." He said the confinement dairy lure was easy to get hooked on when milk was supported at 75 to 80% of parity and there were numerous tax credits and subsidies available.

"All that's gone today. A dairy has to be able to survive on $ll cwt. milk and after-tax dollars. Our present system of dairying is no longer economically feasible because we cannot transfer the capital investment from one generation to another," he said.

Paradigm Shift

An Ohio State researcher told me the major breakthrough at Ohio State this year was the discovery that the grass actually liked to stay still and the cow actually liked to move around. This was a major shift from the paradigm that the cow should be kept still and the grass moved to her in the form of hay, silage or greenchop.

The statistics in the number of dairies and where the growth in milk production has occurred show the zero graze dairy was totally unsuited for humid climates. Most humid states have lost between 60 and 80% of their dairies since the zero graze movement started.

In humid climates the odds are less than one chance in four that a dairy-cow-quality hay can be made. Even with more weather proof silage and haylage the ground is often too wet to get the machinery across when the forage crop is at its prime. To offset the low quality of the forage, dairymen had to feed expensive protein supplements and more grain.

The move to forage cropping and machinery harvest shifted the production advantage from the traditional dairy states to the irrigated areas of the Southwest and West where the absence of rain made forage quality much easier to control. California will soon become America's dairyland unless Wisconsin quickly changes its ways.

The only consistent way to get quality forage into a dairy cow in the humid states is to take the cow to the forage, not the forage to the cow. More and more dairymen are finding it more

51

cost effective to buy their dairy quality hay from the irrigated West, concentrate their efforts on grazing and "salvage" excess growth as dry cow hay or silage.

A major cost breakthrough has been in the perfection of the technique of stockpiling autumn grown pasture and then tightly rationing it out with electric fences to dry cows and replacement heifers.

"It takes two tons of hay to winter a dry cow in Wisconsin," explained Mineral Point, Wisconsin, dairyman, Paul McCarville. "We can get at least a ton of that out of stockpiled pasture."

John Cockrell said that these stockpiled forages can be relatively high in quality. He said stockpiled quack grass at the Charles Opitz farm was still testing 17.3% protein and 159.2 relative feed value at the end of the winter.

"If you'll take your grass into winter in good quality, it will hold its quality through the winter relatively well. That's a real production advantage we have here in the upper Midwest," Cockrell said.

Quality of Life

The ultimate expression of grass dairying is to go seasonal and only make milk when the grass is green. Again, it was the seasonal dairy that made Wisconsin the cheese capital of the United States.

With a seasonal dairy, equipment costs can be exceptionally low. High throughput, New Zealand-style "carport with a pit" milking parlors can be used because the cows are not in lactation in the winter. With a per cow income in excess of $1000 per cow, return per acre can be exceptional and a small farm can provide a very comfortable upper middle class living with a two to four month "off season" that dairymen find extremely rejuvenating.

Carl and Kathy Pulvermacher of Bear Valley, Wisconsin, went seasonal in 1991 and were able to take their first vacation since their honeymoon. Last winter Carl and Kathy went to Mexico and this winter plan to go to New Zealand.

"The nice thing about a grass dairy is that if Carl wants to go off on a trip during the summer, the kids and I can run it fine by ourselves," Kathy said. Last summer Carl went to Russia for a month to work with grass dairymen there.

"My family and I have decided we are not going to work more than 40 hours a week dairying. That means no more than five hours a day. We will not work harder to make more profit. We want the same quality of life as the people have in town," explained Minnesota dairyman Dan French.

By going to grass French could run a lot more cows on the same acreage with less effort. "Grass dairying is a form of farming where it is easier to run 75 cows than 50, and 120 cows than 75."

Going seasonal eliminated the need for a large barn on the McCarville farm. He found his dry cows do fine wintered outside. He feeds roundbales of hay on his pastures during the winter to spread the manure around. **Charles Opitz winters his dry cows and replacement heifers in an open Canadian style feedlot with slatted board windbreaks and mounds.**

Dan Patenaude of Highland, Wisconsin, winters his dry cows by letting them eat from the face of his grass silage stack as is done in Ireland. Then he used large round bales pre-positioned on his paddocks to prevent having to spread manure. The bales are surrounded by an electric fence that prevents the cows from having access to but a portion of a bale at a time. "The only thing I have to do in winter to feed my cows is to move an electric fence," he said.

Dan's wife, Jeanne, said the nice thing about a seasonal, grass dairy was that the whole family could live solely off the farm. With a high margin grass dairy, off-farm work becomes more a choice than a necessity. "I like my off-farm job as a school nurse, but it's nice to know I could quit if I wanted to."

Taste of the Good Life

"I can make a good living on 30 Jersey cows and there is absolutely no way a confinement dairyman in California or anywhere else can put me out of business," bragged Wisconsin dairyman Mike Cannell. Cannell said that **anyone who sits down and figures the cost of making milk in the winter versus what you get for it will go to seasonal production.**

Cannell is like many seasonal dairymen. Once they have tasted the good life they become extremely loathe to do onerous tasks. For example, he no longer uses A-I and breeds his cows natural service to Angus bulls. This means he must buy in his replacements but he sees this as no great loss. "Put a pencil to it. You'll make far more money per acre of grass by milking more cows and buying in your replacements," he said.

Cannell and his wife decided they would try to make their dairy as low labor as possible. "I've decided that I only want to milk cows for an hour and half a day and that is all I'm going to do," he said.

"Now keep in mind that the New Zealanders will milk a hundred cows or more in an hour and a half with their high throughput milking parlors," John Cockrell pointed out.

Most seasonal dairymen are finding that their milking facilities soon become a major constraint on growth. Carl Pulvermacher used to scoff at the idea of a milking parlor but no longer. "When I started out in dairying," Carl said, "my idea of an ideal dairy was one with a big barn, four or five silos and a small house. Now, I would like to have a big house, no barn, no silos, and one of those high throughput, carport and a pit milking parlor."

Carl is wrestling with the decision of whether to retrofit his farm or sell it and buy a non-dairy farm and build it from the ground up as a seasonal grass dairy. "I can smell the future coming and it smells like money," he said. **Entrepreneurs are going to make a lot of money in the Midwest in the next few years with grass dairying. I'd like to be one of them.**"

John Cockrell agreed but said the transition from a traditional dairy to a grass based one is not easy due to the steep learning curve it requires.

"It's tough. It's not simple. You have to think about what you are doing. You're going to make some mistakes along the way. Your pastures are not going to be up to speed for several years, so you need to be willing to suffer financially in the short term to learn this new way of dairying. Keep in mind **if you are not willing to see your job as primarily managing grass rather than milking cows, I would urge you not to even start.**"

Chapter 11

God's Own Wintergrazing

Cameron Parish, Louisiana, is said to be the only county (called parishes in Louisiana) in the United States without a railroad, an incorporated town or a stoplight. With the exception of a single two-laned loop of paved highway, boats and horses are still the primary means of intra-parish transportation.

The marshy land runs as flat as a table-top from the black soils of the Louisiana prairie south to the Gulf of Mexico. It is a treacherous topography and only a trained eye can see where the land ends and water begins.

Alligators and poisonous snakes by the thousands await the unwary and mosquitoes swarm so thick on summer nights they can kill cattle by clogging the animal's nostrils so tight it can't breathe. Hardly the kind of country you would think of as prime cattle country, right?

Wrong!

Cameron Parish is one of America's prime cow wintering areas. Each fall some 40,000 beef cows are driven down from their leased summer ranges in the Piney Woods and turned out to winter on the nutritious regrowth of burned brackish marsh grasses.

With a per head lease cost of less than a dollar per cow per month, the hardy marsh-born cows winter in the marsh from mid-October to mid-April with no supplemental hay, feed, salt or minerals. (The brackish water provides all necessary salt and minerals.)

Calving In The Marsh

The cows are expected to calve unassisted in the marsh and bring at least a 200-lb. calf to the weaning pens in early April. The calves are short-weaned and sold at that time and the dry cows turned out on leased timber company land in the Piney Woods for the summer. Timber grazing leases range from $3 to $7 a cow for an April to October grazing season.

The marsh graziers feel that short-weaning at three to four months of age allows the cow to stay in better condition and thus have a higher rebreeding percentage. The huge sales of these hardy, light marsh calves each spring attract buyers from all over the South and West.

With a total cash cow cost (including land) per annum of less than $50, this marsh/woodland range program is probably one of the most profitable cow-calf programs year-in and year-out in the United States, retired range conservationist Bruce Lehto said.

A major wintergrazing lessor is the Sabine National Wildlife Refuge. The 142,000 acre bird refuge offers a winter home for 1000 cows run under three permits. The lease costs 85 cents per cow per month and runs from October 15th to April 15th.

"A lot of people in other areas criticize us for having such a low fee, but we feel the cows are essential to providing a good bird habitat," explained Sabine Refuge Project Leader, John Walther.

The marsh is burned each winter to remove old growth and to promote the growth of tender winter annuals but Walther explained that the marsh grass grows so fast that it is impossible to keep it short and tender enough for the ducks and geese to use without grazing.

56

"Grazing helps us sustain the benefits of a burn," he said.

Walther said that the marsh grass not only sustained cattle and birds but was essential in the growth of shrimp. The young shrimp spend the growth part of their lives in the marsh feeding on decaying grass.

"No grass, no shrimp. It's that simple," he said.

While most of the residents now have French surnames from intermarriage with the Acadians who settled the Louisiana prairie, most marsh graziers are direct descendants of English seamen who shipwrecked on the cheniers in the 1700's. During the Civil War, Cameron Parish was called "Wild Cow Range" and was all open range until just a few years ago.

"Every morning there would be 60 to 100 cows on the lawn of the local Catholic Church," Walther recalled.

Grazier J.B. Meaux said that grazing beef cows in the marsh has provided his family's primary income since they settled in the tiny community of Grand Chenier in the early 1800's. Chenier, French for Oak Ridge, are the remnants of old beaches that once marked the Gulf's edge and are the only high land in the parish. It is on these cheniers that Meaux and most of the parish's other residents live.

Meaux has been recognized by the Society of Range Management and the Soil Conservation Service for the outstanding job he has done as a marsh grazier.

Cameron Parish has been slowly built up over hundreds of years from sediment drifting westward from the mouth of the Mississippi River. Unlike Louisiana marshes farther east, the Cameron marsh has solidified enough that cows can walk on it-- with difficulty.

To facilitate his cows' movement around the marsh and to even out the grazing pressure, Meaux and other graziers have built dirt walkways out into the marsh. The cows graze outward from these walkways and then return to the walkway to ruminate and sleep. Young calves also stay on these walkways and do not follow their mothers into the alligator infested marshes.

Care is taken to alternate the burrow ditches alongside the walkway to prevent salt-water from invading the brackish marsh and upsetting the fragile marsh ecosystem. Meaux uses Hereford and Limousin bulls and has found the F-1 Brahman/Holstein cow to be the best marsh grazer. **Cows used in the marsh have to have been born in the marsh to not be afraid of it. This also applies to cow ponies.** "We have to be careful to only use marsh ponies in the marsh. A dryland horse tries to fight the mud and will eventually panic and hurt himself, but a marsh pony just takes his time and picks up one foot at a time," he said.

Meaux still prefers a horse for transportation but his youngest son, Mike, has found a wide-tired 4-wheeler a much faster way to get around the marsh. Meaux said both of his sons planned to be graziers after college.

Until recently Meaux grazed 700 cows on the marsh but has had to cut back to 250 cows because he lost his summer woodland range lease. Hard times in the soybean business have shifted some of the prairie country that lies just north of the marsh back to pasture. He now summers his cows on the prairie rather than in the piney woods north of the prairie.

He brings his cows into the marsh in mid-November and moves them back to the prairie in April. This annual transhumance allows the cows to miss both the mosquito season and the hurricane season. His family at one time used to summer cows in the marsh, but they lost 500 cows in Hurricane Audrey.

    Meaux said he had seen the cows start to circle (<u>rôder</u> in Louisiana French) to escape the mosquitoes in the summer and continue to circle until they buried themselves so deep in the mud they couldn't move.

    "It's better for the cows and the grass to get the cows out of the marsh in the summer," he said.

    Until recently, Meaux drove his cows to their summer home in old-fashioned overland three-day cattle drives but he had to stop it due to increasing highway traffic. He now trucks them.

    When he was a boy his family had a 100 mile cross-marsh drive and they had to swim eight to ten rivers and bayous. The drives had to be timed carefully so the cows would hit the rivers at high tide to prevent bogging up on the banks. The family's picturesque cattle drive was even featured in an issue of the **National Geographic** magazine.

    Meaux still carries the traditional symbol of the Southern grazier, a long "cracker" whip coiled around his saddlehorn. He admits he misses the old trail drive days. (The word "cracker" was originally a term for Southern cowboys due to their long "cracking" whips but was extended by Northerners to include all rural Southerners.)

He said horses, dogs and working cattle are the primary social life in the parish and that one only had to announce he was working cows and a dozen neighbors would show up on horseback to help.

There are still quite a few intra-marsh cattle drives each fall and spring that attract droving enthusiasts from all over Louisiana and Texas. Head drover and coordinator for these drives is parish sheriff James "Savo" Savoie.

Unfortunately, there is a growing shadow over this cowboy heaven. Man, in his infinite wisdom, has so tightly diked the Mississippi River that it now drops its silt load off the continental shelf and no longer builds marshlands in western Louisiana. This is allowing Cameron Parish and much of South Louisiana to slowly sink beneath the ocean.

"Every year we see salt water where we didn't see it before," Meaux said. "Every year the Gulf creeps a little closer."

Louisiana is now losing some 60,000 square miles of itself each year to land subsidence. This bodes ill not only for the graziers but the whole Gulf of Mexico ocean food chain. It is the unique combination of organic matter from the Mississippi River and the presence of these brackish marshes that makes the Gulf the most productive fishery in the world.

Recently the Corps of Engineers agreed to start spilling the Mississippi River through the marshes during high water to try and stabilize the situation.

Chapter 12

Cows and Clover Seed Replace Soybeans on Mississippi Plantation

The southwest corner of the state of Mississippi, known as the Natchez District, fits the stereotypical image most people have of the Deep South.

Giant oaks hung with Spanish moss frame white columned antebellum mansions by the score. Port Gibson, a town declared by General U.S. Grant as "too beautiful to burn" still has her good looks and Natchez looks like, and frequently serves as, an 1850 movie set.

The Natchez District has for a long time not been a land of cotton. It did take a flyer at becoming the land of soybeans in the 1970's with disastrous results. The loess soil of the region washes like sugar when laid bare to the rain and the perpetually hot August sun shrivels soybean yields to half those of the Midwest.

It is a land best kept in grass and trees, but seemingly this is a lesson that each generation must learn for itself.

On Chicago Plantation at Hermanville, operated by Jeff and John Seagrest near Port Gibson, the lesson has been learned. From an all-time high of 3500 acres of soybeans, Chicago has none today. "We quite beans five years ago," Jeff Seagrest said. "It was kind of embarrassing to be building fences back in the exact places we had torn them out ten years before, but we did it."

Half the land at Chicago is now in timber and the other half is in intensively managed pasture. The only cash crop Chicago grows today is clover and grass seed.

"We should have done it (made the conversion) three or four years before we did," Jeff Seagrest said. He admitted "heavy metal disease" was very nearly fatal for the plantation in the early 1980's.

"The last combine we bought cost more than the entire plantation did when we bought it in 1950. When we finally decided to pull the plug on soybeans we had 17 tractors in the field. Today we only use two."

Seagrest said luckily the plantation had never gone entirely cash crop and had kept a beef cow herd, a winter pasture stocker operation, a pig finishing operation, some timberland and a custom logging operation.

Chicago In Mississippi

This spirit of diversification has always been a part of the plantation's heritage and even earned it its name Chicago.

The previous owner had a small sawmill on the plantation and during the depression when the market for lumber evaporated began to build numerous barns and out buildings as a way to keep the plantation workers busy.

The neighbors watching all of this construction going on in the middle of the depression finally sent a delegation to the planter to ask what in blazes he thought he was doing. "Hell, I'm building Chicago," the planter replied, and the name stuck.

"We sat down and did an analysis of all of our enterprises and found that year in and year out, beef cattle were our best enterprise. By shifting back and forth between cow-calf and stockers, we think we can stay in the black ten years out of ten." As an example of this flexible approach, they not only discontinued their purchased stocker program but did not even stocker their own steer calves in 1990.

62

"(In 1988) the market paid us 38 cents (value of gain) to wintergraze our steers, but would have paid us a dollar a pound to have sold them in the fall as weaners. In 1989 when weaners were again a dollar a pound, we let the market have them." It was the first year in the previous ten that he had sold steer calves as weaners rather than yearlings. Conversely, his value of gain analysis indicated that winter stockering his heifers was still a good enterprise.

"In 1989 fall heifer calves sold for $14 a hundred less than steers, but yearling heifers in May brought within 25 cents a hundred of what steers brought. If you'll do a value of gain analysis, the market will show you what you should do." Some of his best money has been made grazing slaughter bulls and thin cows and he always watches for that kind of opportunity.

"It seems the people who complain the most about beef cattle are the ones who stay with one enterprise year in and year out, good times and bad. With a grass program you don't have to do that. You need to stay flexible."

The plantation in 1990 had 525 Brahman-cross cows and was building toward a base herd of around 800 cows. Two big cost cutters for their cow-calf program have been calving later in the spring and intensive grazing. "I don't know if we'll ever go back to a purchased stocker program again. With intensive grazing, we can produce a calf for less than a hundred dollars. I doubt if we'll ever be able to buy them for less than that."

Drought Spurs Intensive Grazing

The calving season has been carefully timed to mesh with the start of the spring ryegrass lush. During calving all paddock gates are opened and continuous grazing is used.

Chicago's grass program is a base of improved bermudagrass oversown with annual ryegrass and clover in the fall for the cows and calves and prepared seedbed annual ryegrass and clover followed by volunteer crabgrass in the summer for the stocker cattle.

All but one pasture has been subdivided into eight to fourteen paddock cells. Jeff said he didn't get really serious about intensive grazing until a severe spring drought. "We were actually feeding hay in June that year. I called Steve Pittman at Mississippi Serum and said Steve, let's build fence. We put in a 14 paddock cell that allowed us to not only get the cows through the drought, but we actually we were able to cut hay from it later in the summer."

Jeff said a 14 paddock bermudagrass cell can pasture a beef cow and her calf all year on slightly more than one acre and can in a pinch support a cow and a calf per three-quarters acre during the summer bermudagrass season, and still produce a surplus of hay.

Timing Pasture Shifts

"I like a 14 paddock cell for bermudagrass as I've found our correct rest period for bermudagrass is around 14 days. With a one-day move that's the rest period I get. The whole trick to grazing bermudagrass is to keep it young and tender." He times his moves so that the bermudagrass is only three or four inches in height when he turns the cows on a fresh paddock.

"If I look out there and see that my forward paddocks are getting up around four to five inches in height, I start dropping out paddocks for haying. You've got to keep bermudagrass young for it to be any good."

While many people curse bermudagrass because of this fast maturity cycle, Jeff loves it. "There's three things that have kept us in the cattle business--improved bermudagrass, big roundbales, and electric fences. If I had not had any one of the three, I'd have said the heck with it and just leased the whole place out a few years ago. **The electric fences are worth what they cost just in making your cattle docile and easy to handle. I bet they've paid for themselves in less shrink."**

Jeff is very proud of what he calls his "permanent" winter pastures. He does not put out new clover and annual ryegrass seed each fall, but merely disks the land lightly in the early fall and lets the plants volunteer from seed droppings the previous spring. He showed me a pasture that was half volunteer and half newly seeded and I could tell no difference whatsoever.

"The only difference is that my permanent winter pastures cost me about ten dollars an acre and everyone else's cost them about $70 an acre," he said.

The key to these low-cost winter annual pastures was Seagrest's rediscovery of an almost forgotten clover known as Ball. Ball looks like white clover but is an annual. The Seagrests found that Ball reseeded itself extremely well year after year under the "permanent winter pasture" regimen they had developed. In fact, they were so impressed with it they became a major seed producer.

Today they grow some 200 acres of Ball clover for seed production. The clover is grazed until April and then allowed to flower and set seed, which is cut and combined in June.

"Clover seed production can be a nice sideline but it is very weather sensitive. You need three sunny days in a row for the clover to dry enough to thresh. In 1989, we didn't thresh one pound of clover seed due to the heavy spring rains."

Honey bees are a key to clover seed yields and Jeff has an arrangement with a bee keeper to bring in bees during flowering. This has doubled his clover seed yields. For normal reseeding removing cattle from Ball clover is not necessary.

The Seagrests also grow Bahia grass and Crimson clover seed.

Another profitable by-product of their clover-rich pastures has been a tremendous bounty of deer and turkeys. Hunting leases are now a major source of income to the plantation.

Old times, while not forgotten, are changing down in the Natchez District. Cows and clover seed may not have the cachet of cotton or even soybeans, but they have kept one Mississippi plantation from becoming "gone with the wind."

Chapter 13

Back from the Brink

In the early 1980's with the Midwest farm economy mired in an ever-worsening depression, Ron McBee realized that he was going to have to get serious about making his small farm pay for itself or he was going to lose it.

"I was like most people. I had paid way too much for the land and it was threatening to financially drag down everything else in my life," he said.

The first thing he did was replace his small cow-calf herd with a seasonal stocker operation on his farm near St. Joseph, Missouri. In 1983 he ran 35 heifers on his 60 acres of 96% fungus infected fescue for 150 days. This produced a gain per acre of 88 lbs. of beef--better than cow-calf but far short of paying the land note.

In 1984, he subdivided his farm into 12 paddocks and cut the fescue with red and white clover. He stocked his farm at 120 heifers for 104 days and this produced a gain of 148 lbs. per acre and allowed a surplus hay cutting of 1.4 tons per acre.

In 1985, he subdivided his land further into 31 paddocks and stocked it at a rate of 150 heifers. This was a one to two day rotation per paddock. McBee estimated that in 1985 his paddocks had diversified into a plant mix that was only 50% fescue, the remainder being legumes, Kentucky bluegrass and warm season natives. This pasture salad in combination with the one and two day moves produced a gain per acre of 424 lbs. and the wolf at the door suddenly began to not howl so fiercely.

In 1986, he began to experiment with adding alfalfa to his pastures and broadcast in a spreading variety of alfalfa in four paddocks. The next year he broadcast an erect variety of alfalfa in four paddocks and no-tilled paddocks as a test with borrowed equipment. "I don't know which method worked best. Both broadcasting and no-tilling produced an alfalfa stand for me. I would use the upright variety rather than the spreading variety though," he said.

By 1988, McBee's pastures were estimated to be down to a 40% infected fescue percentage. Alfalfa made up 30% of the stand, clovers were 20%, and warm season natives were 10%. McBee then felt confident enough with his intensive grazing program that he bought the neighboring farm of 270 acres and subdivided it into a 27 paddock cell.

Skinny Versus Square Paddocks

"On my first farm I used a central watering lane with long and skinny paddocks radiating out from it, but on my new farm I designed it so the paddocks were almost perfectly square. A New Zealander visited me and pointed out that **the long and skinny paddocks had the cattle walking over their feed much more than a square paddock system**," he said.

McBee's project now is to get water in each paddock so that he can go to a leader-follower system. He has gotten back into cow-calf and plans to use them behind the stocker cattle. The new farm has also been seeded into clovers and alfalfa and McBee credits the alfalfa with a virtual doubling of his stocker cattle's average daily gain.

For the last several years, McBee has retained ownership of his cattle through the feedlot and has found this much more satisfactory than dealing with the vagaries of the feeder cattle market. In some years McBee ran gain cattle for his custom feeder to keep his stock numbers up but now prefers to own all the cattle himself.

"My program is to buy a heifer in the spring, put 250 to 300 lbs. of cheap gain on her on grass and then take her through the feedlot in the winter. This is a program I have found that works for me and will allow me to buy and pay for land with cattle. As long as it keeps working, I'm going to keep buying farms and getting bigger," he said.

Chapter 14

Grass Dairying in Texas

In the frosted, yellow-brown pastures of the East Texas winterscape, Steve and Pam Roth's dairy in Grand Saline stands out like a Rembrandt painting on a whitewashed wall. Here the pastures are bright green in ryegrass, rye and clover, accented with tightly bunched black and white cows. It looks like a grass farmer's dream and that's what Steve Roth calls it--"a real dream."

It is somewhat ironic that perhaps this best example of grass-based dairying in Texas has not been put together by a local grazier but by a sun-seeking ex-Iowa dairyman.

"With a conventional dairy, the Midwest has the edge because the feed is cheaper," Steve said. "The real advantage the South has is in grass-based dairying because of the ability to grow grass year-round."

Steve credits his move South primarily to his hatred of cold weather and machinery.

"You know how some farmers just love machinery? Well, I'm the other way. I hate it. I wanted to move somewhere the cows could gather their own feed year-round," he said. "I got sick of the cold and I got sick of corn silage."

Steve and Pam graze 500 Holsteins on 230 acres of pasture. The pasture is all Coastal bermuda overseeded in the fall with winter active Elbon cereal rye, ryegrass, Crimson and Arrowleaf clover.

Following his anti-machinery bias, **Steve has only one tractor, grows no crops, and owns no hay or tillage equipment. All pasture overseeding and haying are done on a custom basis.** "I pay a contractor nine dollars an acre to drill in cereal rye and that includes the cost of the seed," he said.

Most of the hay and all of the feed are purchased off the farm. The dairy buys all of its replacements and breeds most of its cows to beef breed bulls. Natural service.

Steve found there was an excellent demand for Holstein/beef cross calves in his area. In 1991 he got $150 for a day old beef cross versus $110 for straight Holstein.

The farm is divided into 42 permanent paddocks made up of four acre strips. Each of these paddocks has its own 1000 gallon round concrete tank. These paddocks are further subdivided in half or into thirds with temporary electric fence, depending upon the rest period needed for the grass.

The cattle are grazed leader-follower with lactating cows top grazing the paddock and then gleaned by dry cows. Lactating cows are given a fresh paddock daily, sometimes twice a day.

Most of the cows start lactating in October and November to avoid the extreme heat and poor quality pastures of late summer. The herd is split into three winter lactating units with one unit dry. Each unit is 100 to 160 cows. Ideally, he likes a 160 cow unit and a stock density of 40 cows per acre.

Midwest and South Compliment Each Other

"Down South, your cheapest milk is made on winter pasture. In the Midwest it is made on summer pasture. The two regions are very complimentary to each other."

Cows producing over 40 lbs. of milk a day are fed eight pounds of a ration consisting of whole cottonseed, whole oats and cracked corn. The whole cottonseed helps to keep butterfat high at

70

3.7%. All lactating cows receive a commercial dairy ration in the milking parlor. Free choice Coastal bermuda hay is kept available year-round to keep pasture dry matter high.

The cows are never shedded. All hay and supplemental feed are fed on the paddocks. The hay feeders and feed bunks have wheels and are shifted with the cattle through the paddocks in a coupled train. The lactating cows are given first choice to pick over a hay feeder wagon. The leavings are then fed to the dry cows. Each hay feeder wagon holds five large round bales.

"By going to rolling hay feeder wagons we figure we have cut our hay waste by two thirds," Steve said. **"Also, we don't get boggy places around the feeders and the fertility is more evenly spread over the paddock."**

Midwestern Versus Southern Replacements

He has found there is a lot of difference between buying Midwestern replacements and Southern replacements in heat tolerance and pasturability. "The Midwestern cows are too fat and are not heat acclimated. Holsteins from Arkansas and Southern Missouri will get out and graze in the heat of day just like beef cattle."

Steve believes Holsteins' poor summer grazing reputation probably reflected pasture management more than genetics. Unlike summer pasture management in the Midwest where growing enough grass was the problem, he found the problem in the South was keeping ahead of the fast growing Coastal bermuda grass.

"We don't want the Coastal to ever get over four inches in height before we enter a paddock. **By keeping Coastal short and vegetative we've found we can make more milk than with alfalfa hay,"** he said.

Thanks to neighboring County Agent Johnny Cates, Steve became acquainted with Dr. Bill Oliver's work in Louisiana on grazing Coastal with stocker cattle. Steve adapted Oliver's techniques of short, highly fertilized Coastal to dairy cows. "Johnny Cates said I would never be able to manage my Coastal correctly without shredding my pastures after each grazing, but I still don't own a pasture shredder. With intensive grazing I can let the dry cows do the pasture shredding for me."

Oliver's research showed the need for frequent nitrogen applications to keep the Coastal's protein level high and the grass young and vegetative. Oliver had recommended a fertilizer application every 28 days during the growing season. Steve modified this recommendation slightly. "By studying our milk production we found there was a sharp drop off after 21 days and now fertilize with 80 units of N every 21 rather than 28 days (every 40 days on winter pasture)."

Thanks to the high stock density and feeding on the paddocks, P and K has increased each year without fertilization. With intensive pasture management and supplemental grain feeding he is able to hold milk production at 50 lbs. per cow year-round for a total feed cost of $4 to $5 cwt.

"That summer milk doesn't come cheap in the South, however. It takes a lot of grain to keep production up. As I said, **the real opportunity in the South is maximizing the use of winter pastures and minimizing late summer and early fall lactations.**" Steve said his milk production costs were a third less in the winter than in the summer.

Other than cold weather and machinery, Steve's other big hatred is milking cows.

"I don't think a man can have a quality life having to milk cows every day. I have found that if you concentrate all of your energy and attention on producing quality pasture, you'll have plenty of money to hire the cow milking done."

He, Pam and his two sons manage the grass and cows, and three hired men handle the milking. Avid horsemen, the Roths bring the cows up from the paddocks with horses.

Design For Ease Of Maintenance

All waste water from the milking shed is captured in a lagoon and spread on the paddocks with a reel-type hose irrigator.

"When I laid this farm out, I put the storage lagoon at the highest point so I could gravity irrigate out of it. I have enough capacity that I can store my waste water in the wet winter and spring months for use on the pastures in the dry summer and fall."

Because the cows are never shedded there is no manure to spread and the waste water is mostly grey water and can be spread on the pasture with no problems of pasture refusal by the cows.

Steve said water quality regulations are already onerous in East Texas and are expected to get worse. These regulations are expected to put many confinement dairies out of business.

As in most grazing operations, your profit is made primarily by what you **don't** spend.

Steve's total investment in land, equipment, fencing, water reticulation, his home, his cars and trucks, everything totaled less than $1000 per cow. He said a comparable confinement dairy would require $4000 a cow in capital costs.

"I figure a dairyman can leverage 75 percent of the total cost of a grass dairy and easily pay it out in five years while making a good living." To prove his point, he is putting in another 500 cow grass dairy for his son near Tyler, Texas.

"Take it from me," Steve said, "If you'll hire the milking done, grass dairying can be just like being on holiday."

(Update: The Roth family now grazes 1500 cows in East Texas.)

Chapter 15

Climate Offers Dizzy Diversity

Hawaiian ranchers deal daily with a diversity of climates and rainfall totals that would make most of us mainlanders dizzy.

For example, the Ponoholo Ranch on the northwest coast of the island of Hawaii rises from sea level to 4800 feet in elevation in the Kohala Mountains. This change in elevation produces a rainfall differential of over 55 inches a year, and a climate that ranges from a five inch rainfall, cactus-studded desert at sea level to a land of constantly blowing and misting cold rain in the uplands. It is like having part of your ranch in Ireland and part of it in Arizona.

Ponoholo is Hawaiian for Von Holt, the family who has owned the ranch since 1928. Today, the ranch is managed by descendant, Pono Von Holt and his wife, Angie.

Ponoholo grazes both sheep and beef cattle and is vertically integrated from birth to slaughter. Stocker cattle are purchased from smaller ranchers when available and day-old dairy calves are also purchased and grown out for slaughter. The big island has no organized cattle auction and all stocker cattle are bought and sold by private treaty.

The majority of the beef cows on Ponoholo are black baldies, which are called "hapaholi" or half-breeds in Hawaiian. Hereford, Angus, Brangus, and Charolais bulls are used. Like most Hawaiian ranches, the cows were bred up from straight Herefords, which was the predominant breed in Hawaii until recently.

Since 1982, Ponoholo Ranch has been on a crash program to subdivide its pastureland into grazing cells and today 18 separate cells are in operation. A doubling of the cowherd was completed in 1987 and production costs per pound of beef have fallen by one third since intensification was begun.

However, the ranch has now slowed capital investment and is putting all of its energy into fine tuning the management of its new cells.

"You can build cells faster than you can train people to run them. We are having to slow down and train a whole segment of middle management in grazing cell management," Von Holt said.

Cells In High And Low Rainfall Areas

The ranch has built cells in both the high rainfall upland areas of the ranch and in the very low rainfall desert areas along the coast. However, if he had it to do over again, he would put all of his money and effort into the high rainfall areas and would have continued to run the dry areas as traditional extensive ranches until all the opportunities in the wetter areas had been realized.

"We can get a payback of all intensification costs in the high rainfall areas in six to nine months. We had programmed a payback on investment of five years in the low rainfall areas, but it now looks like seven to nine years just to recover our capital from the low rainfall areas," he said.

Some 3000 acres of 60 inch rainfall upland pasture has been highly subdivided into New Zealand-style paddock blocks. These acres are used primarily to grow out stockers, replacements, and lambs produced on the drier lowland acres.

On the wet uplands, the base forage of Kenyan Kikuyu grass is overseeded with annual ryegrass, rye, and Israeli Haifa

75

white clover to provide high quality winter and spring feed. Ellett perennial ryegrass has been tried but cannot compete with the highly aggressive Kikuyu. The Kikuyu must be grazed into the ground each fall to allow overseeded cool-season grasses and legumes to establish. This "pasture flogging" technique also prevents the mat-like buildup of dead grass and stems common to continuously grazed Kikuyu.

The upland cell can produce eight pounds of beef per day or six pounds of lamb per acre during the 90 day spring lush. Year-round the upland cells generate 1.5 lbs. per head per day stocked at 1.5 to 2 stocker cattle or 1/4 pound per head per day at 15 lambs per acre.

Kikuyu, much like bermudagrass on the mainland, produces acceptable average daily gains with stocker cattle when handled in a leader-follower system, but produces low gains under continuous grazing with only stockers. The Kikuyu is such an aggressive grass that it is frequently seen growing on the tops of fenceposts and has pushed out virtually all of the temperate grasses and clovers that predominated in Hawaii forty years ago.

According to Von Holt, prior to the coming of Kikuyu the lower acres were naturally irrigated by runoff from the constant rainfall at the top of the mountain. Today, the Kikuyu sops up all of the runoff and has turned the lower acres into a desert. Rain only falls on the leeward low elevations in the winter.

The ranch originally used the centered hub style of cell design, but is replacing it with the New Zealand block style of subdivision. Nutrient transfer is obvious in the centered hubs with dark green grass near the hub center and light green to yellow grass at the far reaches of the cell.

Most of the subdivision is with two wire fences made up of a ground and a hot wire. Such minimal fences, while fine with cattle, are unacceptable with sheep.

To avoid fencing expense, the ranch's sheep are set stocked in one heavily fenced area, and the cattle are rotated through the sheep to keep the grass under control. Such "put and take" methods are necessary to keep the Kikuyu in a quality state for the ewes. Like most Hawaiian ranches, Ponoholo has no haying or silage equipment that could be used in lieu of cows to maintain pasture quality.

The sheep are being bred up from Hawaiian native stock with imported New Zealand Coopworth rams. These crossbreeds are then bred to Suffolks as a terminal sire. The sheep breed twice a year and the ranch sells grassfat lambs to the local market each week.

"We, like most cowmen, thought sheep were somehow more fragile than cattle, but we haven't found that to be the case. While it's true lambs respond to good management and good grass, we run the ewes just like our beef cows," he said.

In terms of net per acre, the sheep are much more profitable than cattle and all future expansion in livestock numbers will be in sheep rather than cattle, he said.

Calving Seasons Match Grass Seasons

The beef cows are split into two calving seasons to match the two major grass seasons on the ranch. The cows on the dry, low acres calve December through March due to the mainly winter rainfall. Cows in the cool, high rainfall, upland, grass area calve in April and May to match the lush summer growing season.

The ranch has a management target of growing all yearlings

to at least 850 lbs. before sending them to the feedlot, but this is not always possible due to rainfall variation and market considerations. "We Hawaiian ranchers have to be careful to spread our production out over the year to avoid flooding the local market. We are committed to sending 300 feeder cattle to the feedlot each month and sometimes we have to ship them out lighter than we would like," he said.

The ranch's feeder cattle are loaded into special containers and shipped by barge to the Parker Ranch's feedlot on the island of Oahu, some 200 miles from the island of Hawaii. Most of the state's beef is born on the island of Hawaii, but consumed on the populous island of Oahu.

The Hawaiian islands produce little grain and must import Canadian and New Zealand barley and wheat for animal feed. A U.S. law that requires interstate commerce on American ships makes U.S. grown grain uncompetitive in Hawaii due to the much higher freight rate of American ships.

To cut overhead costs, Ponoholo Ranch shares office space and workers with neighboring Kahua Ranch. The two ranches partner so closely that many outsiders think of them as one production unit.

Pono's wife, Angie, is in charge of growing out day-old dairy bull calves and orphans from the range. Angie has a New Zealand "Mechanical Cow" that automatically mixes up dry milk powder into warm milk and dispenses it upon demand by the calves. The calves are grown to four months of age on this "cow," ryegrass and clover, and a grain feed supplement.

The Holstein bull calves were originally going to be grown out as intact grassfat bulls as in New Zealand, but they proved to be a management headache around the cows. The calves are now steered and sent to the feedlot when they weigh 600 pounds.

Fewer orphans are being brought to Angie now since they are finding the **orphans are able to steal enough milk on the range to develop normally thanks to the bunching effect of intensive grazing**.

With nearly six years of intensive grazing experience under his belt, Pono finds the primary problem is not cell engineering or animal performance but training and motivating people.

"With such a high variability in output directly responsible to a worker's skill, a whole new type of pay program is needed. We currently have a program where we split 50/50 the gross dollars produced above the norm. It's tough to pay that bonus when you're short of cash because you've got a lot of cattle in the feedlot or whatever, but you've got to do it. There has to be an incentive for your workers if intensive grazing is going to work at its maximum potential," he said.

Chapter 16

Organic Hogs on Pasture

Jodi and Ron Snyder of Port Chester, Pennsylvania, have found that pasturing pigs can cut capital needs and feed costs in their organic pork production, plus produce a much more desirable carcass.

"We are getting a 72% dressing percent on our pastured pigs versus 64% for those raised in confinement," Jodi said. "This is very important because the organic market does not want to see any fat at all. Even with our natural leanness we trim the meat off all external fat."

Jodi said she felt the "pastured" cachet was an important part of their marketing. They sell their pork all over the United States. "A lot of people see pasturing as a more humane way to raise hogs than confinement. Some people who are normally vegetarians will eat pastured pork."

Jodi said the nice thing about a pastured pig operation was that you could bring a customer to see it and they would still want to buy the meat, whereas, one whiff of a traditional confinement house turned off most customers. The Snyders use a lot of rock phosphate to keep odors down in their confinement finishing house. This has another benefit in that the manure acids help make the rock phosphate more available for the pasture.

The sows run on pasture year around and farrow outdoors in the pasture months. They had tried to provide farrowing hutches for the sows but they preferred to build nests in the blackberries. Sows on pasture averaged 8 and 1/2 pigs per litter, which was exactly the same as when the Snyders farrowed them inside in the winter.

"Allowing the farrowing house to sit empty for half the year is a great way to break the disease cycle," Jodi said. Disease prevention is particularly important to organic farmers trying to stay away from the use of drugs. **It is also important that the pigs' pasture be rotated from year to year to prevent disease and parasite buildup.** A small Quonset hut open to the South and with an open window on the North end provides the only winter shelter for the sows who seem to enjoy getting in a big pile when it is cold.

Feed Rations

The pigs are fed a ration of organic corn and roasted soybeans. Ron said it was on the sows that the big feed reduction came on pasture. The growing pigs seemed to eat just as much on pasture as those on a finishing floor but did gain weight slightly faster.

The pigs are used to glean the Snyder's sweet corn fields after harvest and have done well on headed-out oats. Turnips are sown into the corn fields after harvest for fall and winter grazing. The pigs have access to a free choice mineral ration of diatomaceous earth for worms, kelp meal, 10 lbs. of salt, 10 lbs. of rock phosphate, and 10 lbs. of limestone.

All the Snyders' fences are one-wire electric. "The sows will respect a one-wire electric fence but the pigs won't," Ron said. "To keep the pigs content and prevent them from roaming we always put a couple of sows in the paddock with them."

Pig graziers who are buying in all of their grain have to be careful to finish no more pigs than they can market, Jodi advised.

81

With the high cost of organic grain, a finished pig that winds up being sold through traditional channels loses about eighty dollars a head. Feeder pigs, however, can usually be sold through traditional channels without a loss due to the small amount of grain used in the Snyders' production.

"I wrestle all the time with the question if a man should be in organic pigs if he doesn't raise his own grain, which we don't. But then if I could sell my grain for what I am having to pay for it, I probably would be just as well off to sell the grain and forget the pigs," Ron said.

"Our pasture program has done what we wanted it to do," Jodi said. "It has lowered our feed costs somewhat but more importantly it has allowed us to expand numbers without having to invest in more buildings and equipment."

Chapter 17

Partnering in Alabama

Can died in the wool grass farmers find happiness in the feedlot business? Pete Reynolds, Jr. and Sr. and Norman Stokes are trying to find out.

Pete and "Little Pete" Reynolds are west Alabama prairie ranchers and Norman Stokes in a south Florida rancher. Three years ago they decided to "partner" on a 4000 head, slatted floor feedlot that was for sale near Forkland, Alabama. "Little Pete" admits to having been partial to the purchase because he had worked there under the previous owner, but Stokes was looking for a way to give himself more marketing options for his calves.

"Norman now has the option of grazing his calves to yearlings here in Alabama and selling them, finishing them here, or selling off the teat in Florida," Pete Jr. said. "I guarantee you Norman isn't married to any of them either."

This flexible attitude extends to he and his father's grazing partnership as well. "We'll normally sell half of what we graze as feeder cattle and feed half," he said.

The Reynolds graze a herd of 300 Brangus momma cows plus turn a little over 2000 stocker cattle on grass each year.

"Our stocker program is to try and turn 700 head three times a year. Because we are always back in the market as soon as we sell and the turns overlap somewhat, we normally have a grass stocker inventory of around 1000 head on hand. We'll buy another thousand head of cattle to put directly on feed also," he said.

The Reynolds like to feed primarily heifers and heiferettes and sell their steers as feeder yearlings. Between the two feedlot partners, they keep the feedlot two-thirds to three-quarters full and the remaining pen space is offered to outside customers. For the last three years, the feedlot has run at near 100% capacity. Most of the fat cattle go to John Morrell in Montgomery, Alabama.

Pasture And Grazing

A unique feature of the feedlot is a continuous flush system that pumps the waste water and manure into a holding lagoon. From the lagoon the manure and water can then be pumped onto 300 acres of pasture through walking irrigation guns.

By using the irrigation guns and "Faster Pasture" sudangrass, Reynolds was able to graze 900 yearlings on 100 acres for six weeks in the summer of 1987.

Most of Reynold's stocker pastures are Marshall reseeding ryegrass in the winter and crabgrass and Johnsongrass in the summer. His permanent pastures are primarily fescue/dallis mixes with red clover. The cowherd is useful for its ability to handle properly maintained leased pastures. Pasture leases on "dirty" fescue are extremely reasonable in the west Alabama prairie, according to Pete Jr., and cows with a little care seem to handle the toxic endophyte.

Pete normally receives a cent under the Texas/Kansas average on heifers at Montgomery, and two cents under on steers. Dry grain is shipped in by barge to Demopolis on the Tenn-Tom Waterway. He also receives grain on back-hauled lumber trucks and on the nearby Burlington Northern Railroad where he has a spur and unloading facilities. "If you figure we get paid a little less for our cattle at slaughter and pay a little more for our grain, but we save $25 a head in freight and shrink, we're a dead wash with feeding in Kansas. What we don't have are blizzards and winter wrecks."

Pete has been feeding primarily whole shelled corn and supplement, but a recent "partnering" with a large rowcrop farm has added the option of high moisture corn. Sunbelt Farms has traditionally been in wheat, corn, and soybeans. Sunbelt is owned by Jerome Weinstein and managed by Bill London. The Farm has been using Stokes and Reynolds feedlot as a way to market their corn, and last year built a large bunker silo at the feedlot to enable them to harvest their corn at high moisture and avoid the drying costs.

"The bunker gives us the space for 3000 to 4000 acres of high moisture corn," explained London.

Diversified Program

From strictly a cattle feeding/grain farm, Sunbelt has diversified into stocker grazing. "Our yields in Alabama are much worse than what our climate and rainfall should produce," explained London. "This was always a mystery to me until we turned under some old pasture sod and bang, Midwest yields! We found out that we could double our average yields by turning under sod pastures. After three years, the yields nosedive again as the humus burns up. What we are looking at now is a permanent pasture/crop rotation program."

Sunbelt's stocker program turns enough stocker cattle to keep 700 head on feed at all times. "By grazing and feeding, we've never lost more than $10 a head and have made way over $100 a head profit and that includes paying ourselves 40 cents a pound for our grass gain," he explained. "Cattle are going to play a much larger role in our operation in the future."

Gordon Beckett is another new convert to Alabama grass farming who is feeding cattle with Pete Jr. Beckett is an Australian merchant banker who is developing a 2000 acre stocker operation near Demopolis. He keeps a pen of 110 head on feed at all times with Stokes and Reynolds while he renovates his "dirty fescue" with ryegrass and Faster Pasture before planting it back to Fungus Free fescue.

When his new ranch is up to speed, Beckett plans to turn

1200 stocker cattle twice a year on his pasture mixes.

Reynolds said this idea of turning your inventory several times a year was good for both cash flow and market protection.

"There's no way I would put 2000 cattle on feed at one time, but I wouldn't hesitate to put 200 head on feed a month. We've got to start thinking of our grazing operations in the same way," he said. He and his father's pasture operation has become a continuous stream with the bigger cattle being sorted off and replaced by lighter thinner cattle each week. Pete Jr. makes from one to four local cattle sales a week to keep his feedlot and grazing operation replenished.

"To be honest with you, the cattle business has been real good to me. I've never had to do one thing in my life I didn't really want to do. There's not many men who can say that," Pete Jr. said.

Chapter 18

Balanced Year Round Forage
Aim of This Midwest Grass Farm

George Gates is one of the Midwest's largest growers and wholesalers of native warm season grass seed in the United States. He is planting Fescue and cool season grasses on his farm as fast as he can. This has caused more than a few eyebrows to be raised around Grant City, Missouri. After all, Gates has long been one of the most vociferous proponents for native warmseason grasses. Why the big change?

The big change is that George Gates has diversified from pure grass seed production into the beef cattle business and if you are raising cattle in the very northern edge of Missouri (six miles from Iowa) you are going to need cool season grasses for a year round grazing program. "We are not being disloyal to warm season native grasses we are just facing the reality of our situation," Gates said.

The CRP program caused a boom for native grass seed growers like Gates. Unfortunately, this boom brought a multitude of new people into the native warm season grass business and when the CRP program ended the government inspired seed demand stopped and with the newly enlarged seed supply the native grass seed business collapsed.

Gates is sanguine about his shift to grazing. "I've always wanted to be in the cattle business. It's been a lifelong dream." Native warm season grasses will still play a role on his 3000 acre farm. It will just be a smaller role.

Gates said an all warm season grass program limited a grazier to a short season stocker program like that in the Flint Hills. From his analysis these short season programs are too financially risky because not enough gain per head is produced to absorb the traditional price rollback.

Gates prefers a cow-calf/stocker program where the grazier produces at least part of his own calves. "I like the flexibility of buying calves but I like the financial stability of producing some of them as well," he said.

Of course, cow-calf means year round grazing and is why Gates plants most of his farm to cool season grasses.

"I don't think a Midwestern grazier should have over 25 to 30% of his farm in warm season grasses," Gates ranch manager, Norman Kanak, said.

Gates said they originally thought they could use the dormant warm season grass for wintergrazing the way cattlemen do in the West. However, after one winter they have discontinued that idea. The grass quality was so low and required so much supplemental feeding that it was not cost-effective compared to stockpiled fescue.

Stockpiled Fescue Winter Ace

"Stockpiled fescue is our big ace in the hole here in the Midwest. We've got to maximize our winter use of it," he said.

He quickly emphasized the word **winter**. "**We don't graze fescue in the summer and we don't graze it in the fall because we are allowing it to stockpile then.**" Since his cattle are not on fescue during hot weather, he has planted the more durable high endophyte fescues rather than the endophyte-free varieties.

Gates' cattle are on the native warm season grasses from late May or early June to late August. Grazing can be extended up to mid-September if rotational grazing is used and a high leaf residual (six to eight inches) is left.

"You absolutely, positively have to allow a native warm season grass time in the fall before frost to build its root reserves if you want to keep it," he said.

In the early Autumn the cattle graze orchardgrass and brome and a "new" grass Gates calls "Eastern Meadow Grass." "It's also called Quackgrass around here but few can see its potential as a forage unless we call it something else. We are real excited about Quackgrass," he said. All lowland pastures have been planted to Alsike clover, and red clover is used on the uplands.

Alsike Clover Fills Bare Spots

"A problem with warm season native grasses is that they tend to grow in clumps and leave a lot of bare ground around them. Since we are really in the solar collection business in grass farming, I'm not making any money on that bare ground. The Alsike fills in those bare spots, improves both the tonnage of dry matter and its quality, plus it allows us to graze our warm season pastures earlier in the spring and after frost in the fall," Gates said.

He thinks Midwest graziers should concentrate their warm season grasses on their poorly drained lowlands and their cool season grasses on their better drained uplands, particularly in heavy soil areas like Northwestern Missouri that pug easily. "**Your warm season grasses are your drought savior. Put them down on your lowland areas where the moisture will be best during the summer drought and let your cool season grasses go dormant on the uplands. Then in the cool wet months you'll be off your boggy land and on your best drained land**," he said.

Gates found the Alsike clover did very well in their acid, wet lowland soils where he is concentrating his warm season grasses.

Manager Norman Kanak said he thought native warm season graziers needed to add legumes or possibly a cool season annual to maximize the gain per acre of land. "An acre of warm season grass has got to be competitive with the gain from fescue and from 30 to 40 bushels of soybeans to survive in the Midwest," he said.

Gates plans to experiment with overseeding some of their land with cereal rye this fall to see if they can fill the late winter/early spring hole in their forage flow.

"If we can figure a way to get through late February and March on cereal rye, we can quit the hay business altogether," Gates said.

They have subdivided the farm into 40 acre paddocks but these were still way too big to get the stock densities they wanted. Additional fencing was having to wait for more stockwater reservoirs to be built and piping put in. "We're financing this ranch's development out of its cash flow so we can't go as fast as we might like," he said.

He added that while planning out a complimentary grazing program was relatively easy to do on paper, putting it together in the pasture had proven to be much more difficult.

"This is truly the thinking person's form of farming," Gates said. **"You've got to match forage quality and quantity with either where the cow is in her production cycle or with the class of animal. The animal's needs are changing every day, and the grass is changing every day.** It's sure a lot more difficult than growing seed, but it's sure a lot more fun as well."

Chapter 19

Working with Nature
Key to Adams Ranch

"If you take the nitrogen, hay, and feed out of cattle, and you use cows that can breed and calve with a minimum of labor, there's no way not to make money in cow-calf at today's prices," explained Bud Adams, Chairman of Adams Ranches, a south Florida cattle and citrus operation.

"All the money we've ever made in cattle has been with grass," Adams pointed out, and for the last 35 years he has poured a ton of sweat and thought in how to make his grass crop better. He has developed a year-round legume-based grazing system utilizing cool season and warm season legumes to give his cows a high quality pasture year-round, and his son, Mike, is experimenting with a 23 paddock controlled grazing cell.

While many ranchers and analysts believe vertical integration all the way to retail is an answer for the cowman's cost-price squeeze, Adams prefers to concentrate on his 10,000 momma cows and in reducing costs of production. "If other people can make money putting grain through cattle, I'm happy for them to do it,"

Adams said. Ironically, it was the shift of Florida's markets from local grass-fed slaughter to providing feeder cattle for the growing grain fed markets in the Midwest that prodded Adams into developing a new American breed called Braford and set him off on his quest to improve his grass crop.

"We could get more money for a Hereford cross calf as a feeder than we could get for a three-year-old grass-fat Brahman steer in the early 1950's. So we started bringing in Hereford bulls from the west, but we had horrendous problems with them. I was reading Darwin's book on evolution at the time and got to noticing that the deer on our place were a lot healthier than the cattle. I realized that nature had adapted the deer to south Florida and that the key to our success was to adapt Hereford cattle to south Florida. The result was the Braford," he explained.

The same adaptive process has also worked with his forage crops. Adams said that a fertilizer salesman talked him into using nitrogen on his pastures in the early 1950's, but that the sudden lush caused an outbreak of Armyworms that ate most of the grass and cured him of nitrogen fever.

In the early 1950's, Adams flew 1 to 2 lbs. of Louisiana S-1 clover on all of his ranch land despite everyone warning him that the clover couldn't survive south Florida's wet summers and dry winters.

"Probably 90% of this clover died, but 10% of it survived and produced seed that could survive, and that in turn produced seed that could survive," he explained.

Clover Mix

Today, virtually all of Adams' pastures are covered with adapted white clover, and he is starting the same program with a tropical clover called Aeschynomone.

Mike Adams explained the shift to a companion warm season clover:

"Our white clover plays out in June, but the Aeschynomone comes on in July just as the quality of our grasses start to deteriorate due to the summer slump. The Aeschynomone grows until cold weather and then the white clover takes over again."

The high quality forage in the late summer and early fall

allows the Adamses to keep their calves on the teat longer and still have the cows rebreed. Many south Florida cattlemen wean their calves in July and August as the grass loses its quality. This results in a lightweight calf that is many months away from the winter pasture season in the temperate areas of the south. The Aeschynomone allows the Adamses to keep a calf on the cow until it is 9 to 10 months old and sell it into the teeth of the "grass fever" at the start of the winter pasture season.

"Due to our numbers and our weaning season we are able to get a good fall price," Mike explained.

The calves are sold to a broker in Memphis and are shipped over a four month period. The heavier calves go directly to feedlots and the lighter calves go to winter pasture.

Seedstock production accounts for only 8 to 10% of Adams' cattle revenue, but Bud Adams said it is a growing and important source of income to them.

"A successful seedstock producer has to keep his roots in the commercial cattle business. Only a very few outstanding calves should be chosen as seedstock cattle in anyone's herd," Adams said.

Pasture rotation has always been a part of cattle raising in south Florida as cattle went from the marshy areas to the dry areas in tune with the various grass seasons, but Mike Adams has been experimenting with small·paddock/fast rotation grazing on part of the ranch.

Getting Started

"We started with a four pasture rotation system, because it was easy to get started," Mike explained. "Unfortunately a four pasture system does not give you the time control over the grass and is really a put and take system." Mike was constantly having to add and subtract cattle to match the grass growth.

"My advice to anyone considering controlled grazing is to start with a lot of paddocks. With 23 paddocks we can speed up or slow down the rotation to match the grass growth rather than add or subtract cattle. The more paddocks, the more control you have over matching the cattle to the grass," Mike said.

Adams' cell uses one wire for the paddock subdivisions and three electrified wires for the lane subdivisions. "Wherever you are going to crowd cattle, like lanes, you need more wires, but to keep the paddocks divided I think one wire is fine," he said.

With the cell, Mike can run a cow-calf pair year-round with no hay on 1.8 acres of land. He thinks a 300 cow cell is best as 300 cows can be worked by a cowboy crew in one day. The cells all have lateral gates between the paddocks as well as lane gates at the ends of the paddocks.

Mike is particularly interested in the change in the brush and palmetto composition in the cell. The high stock density emphasizes the browsing ability of the cattle and has resulted in a decrease in wax myrtle and palmetto. "If we can substitute grazing pressure for disking for brush control, it could help us control a lot of costs," Mike said.

"Whatever we can encourage nature to give us for free, we want to take advantage of it. I saw an ad the other day that illustrates what's wrong with the cattle business," Bud Adams said. "In one picture they had a cowboy on a $40 horse watching over $200,000 worth of cattle. This was the past. In the other picture they had a $500,000 helicopter chasing a $200 calf. I don't consider that progress," he said.

Nature Speaks

"Problems are nature's way of trying to tell you something's wrong in your management. The biggest single limiting factor in the cattle business is the 'want to' factor. Most cattlemen can't grow clover because they don't want to grow clover. South Florida is the sorriest land in the world. You've got to want to grow something here and it's the same all over. I've got friends in Montana who can produce a calf almost as cheap as we can here by working with nature. If you want to make money, sell your depreciating iron (tractors), put it in cattle, then get rid of the cattle that won't work under your situation," he said.

While his ranches have come a long way in the last 30 years, Adams is far from satisfied with them. "If you come back down here in 35 years, my grandchildren will be seeding newer and better grasses and legumes and breeding even better adapted cattle. It never stops. You've just got to want to do it."

Chapter 20

A Vision of Grass
Guides Anderson Ranch

When the history of management intensive grazing in North America is written a hundred years from now, Anderson Ranch will make the first chapter. According to grazier historians, the 160,000 acre ranch had the third Savory-type grazing cell ever built in the United States, and the second in Texas. An illustration of the newness of intensive grazing technology in North America is that this pioneer cell was built in 1979.

Anderson Ranch manager, Gary Loftin, also will be prominent in that first chapter. He probably has had more hands-on experience with intensive grazing than any other rancher in Texas. Loftin not only has been involved in the intensification of Anderson Ranch from the beginning, but built and ran Texas's first grazing cell on the McElroy Ranch. Like many pioneering efforts, Loftin recalled it was the pressure of hard times that pushed him into the scary unknown of intensive grazing.

Loftin first heard Allan Savory speak at the International Ranchers School in Arizona and was extremely receptive to Savory's early pitch of "doubling your stocking rate." The McElroy ranch he was a partner in, had just lost a big grazing partner, had just lost a big grazing lease, and the prospect of liquidating their cattle on an extremely depressed cattle market had them searching for any alternative. "We finally concluded we had but two choices; we could build grazing cells or go bankrupt," he said.

Loftin and his partners chose the first course and the cells allowed them to hang in there. With the return of better cattle prices the partnership was liquidated, and Loftin signed on with Anderson Ranch. The subsequent ranch operators tore out the first grazing cells ever built in Texas and reverted the ranch to traditional grazing management.

Ranch History

At Anderson Ranch, Loftin has overseen the building of grazing cells that now cover 83,000 acres of the ranch. Following Allan Savory and Stan Parson's advice, Loftin and Anderson Ranch managing partner, Mike Harrison, sat down and for three days drew out cell designs for the whole ranch before the first cell was built in 1979.

Virtually everything needed for the building of electric fences was impossible to find in 1979. "Thank God for Art Snell," Loftin recalled. "He was a great help in getting us started." (Art Snell of San Antonio, Texas, was a pioneer in popularizing electric fence technology in the U.S.)

The sprawling Anderson Ranch had been put together by a railroad engineer in 1918 and was run exclusively as a stocker operation growing out imported Mexican steers. It was managed as a classic extensive horseback outfit with a chuck wagon and cowboys sleeping under the stars under the late 1940's.

The ranch in 1979, still had only one shipping point for the entire ranch. Cattle had to be trailed long distances to be loaded out. The first grazing cells were built to radiate out from newly constructed shipping corrals. Fences extended a mile and a half outward from each water point, with eight to nine subdivisions.

With the vagaries of buying Mexican cattle increasing in the last few years, the ranch has diversified into cow-calf, both registered and commercial, horses, fee hunting and stockers, both

owned and pastured for others. "This ranch would be much easier to operate as a pure stocker operation, but we believe the cowherd gives us an added degree of financial stability a straight stocker operation doesn't have," Loftin said. Conversely, the stockers give the ranch the flexibility to better match the stocking rate to West Texas' highly variable rainfall and grass growth. Rainfall at Anderson averages a little less than 10 inches a year but can vary greatly to either side of the norm. Severe droughts often hang on for years at a time making a constant stocking rate impossible.

The ranch runs around 2300 animal units, of which 1000 are breeding females. Loftin said this is about half the animal units the ranch could carry with 1988's good grass condition, but extremely high prices for stocker cattle deterred them from adding more cattle.

In 1979, Anderson had an SCS-rated carrying capacity of one Animal Unit per 180 acres, was estimated to be 75 to 80% bare ground and was almost a pure monoculture of black gramma grass. Because the ranch is being subdivided in stages, it is possible to dramatically see succession at work due to the varying ages of the grazing cells. One of the earliest cells now has a stocking rate of one AU per 21 acres and is volunteering a crop of Blue Panic grass, an extremely high quality cousin of Johnsongrass. "Can you imagine what you could do with stocker cattle on 8000 acres of Blue Panic?" Loftin asked.

All of the Anderson Ranch grazing cells are of the centered hub design with radial electric fences and incorporate every type of cell center imaginable. The ranch is dotted with many producing oil wells, but the ranch does not own the mineral rights, so their only benefit to the ranch is as an extremely cheap source of pipe and sandline cable for corrals and cell centers. **During droughts when long rest periods become essential, various herds are combined and several cells are run as one huge cell.**

Non-subdivided areas of the ranch are not continuously grazed, but are grazed like a big paddock with a large number of cattle concentrated to graze the area as quickly as possible and then removed. The low pumping rate of the ranch's windmills and the lack of water reservoirs and tank capacity is a primary constraint to increasing stock density using this simple method of time-controlled grazing. Cowboy labor costs to find and remove the cattle are also much higher than with the cell grazing, but the grazing cells have not totally removed the need for a surefooted horse at Anderson Ranch with its one-thousand-acre size paddocks. "Finding a thousand steers on a thousand acres is a lot easier than finding them on a hundred thousand acres but there is still a need for a horse and a cowboy's skills even with intensive grazing in West Texas," he said.

Loftin said that watching plant succession work is extremely scary for a new grazier because it appears the range is deteriorating in the early stages due to the large number of weeds that come in, some of which are poisonous. "Don't panic. This is nature's scab. It's nature's way of healing itself. You have to stick to your plan and don't bail out," he said.

Working With Nature

From their experience, Loftin said that succession is speeded by a high stock density, a long rest period and rain. Succession moves forward almost imperceptibly during droughts but then rushes forward during the rainy periods. The early pioneer weeds of perennial broomweed, thread leaf, groundsels and creosotebush

98

have all naturally died out in the older cells as succession has progressed and grass has covered the bare ground. Loftin believes the mesquite will also eventually die out naturally as the grass thickens.

Loftin believes a higher and higher stock density is necessary to keep pushing succession forward. **Once the animals have plenty to eat in a paddock they resume selective grazing and effectively stop succession.** To get a higher stock density the paddock must be further subdivided. The increase in subdivisions will allow a longer rest period.

He feels a 120 day rest period moves succession faster than a shorter rest period in the dry climate of West Texas. These long rest periods also allow high quality winter legumes and annuals to volunteer in and provide a quality winter range that will allow the ending of supplemental feeding of cottonseed cake. In 1988, annual cash out-of-pocket costs for feed and mineral averaged $35 a cow.

The ranch does not supplement any of their replacement heifers to help genetically adapt the herd for management's ultimate goal of a non-supplemented cowherd. These non-supplemented heifers have been able to produce a 75% average conception rate from range alone.

The ranch has split calving seasons to even out cash flow and market risk. All ranch-born steers are owned all the way to slaughter and are placed on feed when they reach 775 to 800 lbs. Purchased stocker cattle are either sold as feeders or fed out depending upon the market. The range-grown steers leave the ranch with a lot of unexpressed compensatory gain in them. High on Loftin's wish list is some acreage of irrigated land that could be planted in high-quality annuals to "pop out" their steers before sending them to the feedlots.

However, the agronomic and animal benefits of intensive grazing are just two of many benefits according to Loftin. **The most overlooked benefit of intensification is the tremendous lowering of labor costs per animal unit that occurs.**

"Before we subdivided, it took 16 cowboys on horseback days to round up our cattle. I once figured it cost us $16 worth of labor just to brand one calf. Today, we run 2300 head with just myself and two others," he said.

The ranch raises their own bulls from a herd of registered

Beefmasters and raises their own working horses as well. Both of these enterprises have turned into small profit centers as the surplus animals are sold to other ranchers, primarily in Mexico and the Southeast.

Hunting has become another profit center for the ranch. Birds are the primary draw but selling coyote hunting rights has been successful and has provided a way of turning a major nuisance into a cash flow.

Like most pioneers, Loftin and Harrison have had to endure a lot of ridicule from their neighbors and various government agencies. After a period of trying to evangelize their neighbors and the SCS about the benefits of intensive grazing, they have concluded that none are more blind that those who refuse to see. While they still welcome the serious rancher who is involved in and understands intensive grazing, the welcome mat is no longer out for those merely seeking an argument.

Loftin and Harrison admit to being driven by a vision that many of their neighbors find totally ludicrous. Their goal is to eventually be able to ride over Anderson Ranch in stirrup deep bunchgrass just like the first settlers to West Texas were able to do one hundred years ago.

"We can see that vision getting closer every day," Loftin explained. "I believe the time will come when there will be a lot of money in healing ranchland and teaching people how to keep from degrading it again, but for now we're content to just manage Anderson as best we know how."

Chapter 21

Midwestern Ranchers Find Southern Grazing Connection Profitable

Can an upper Midwestern grazier ever find true winter happiness? Can he enjoy watching his calves gambol about on lush green ryegrass pastures in January? And can he come out of the winter with a profit from the experience rather than a cost?

Wisconsin stocker graziers, Reid Ludlow and Larry Smith, are just two of a growing group of Northern graziers who have found they would rather switch regions than fight Mother Nature. Each winter when the snow drifts deep in Biroqua, Wisconsin, Ludlow and Smith's cattle are grazing green ryegrass pastures in the Deep South.

Both Ludlow and Smith buy very light calves (225 to 250 lbs.) in the fall and early winter. These calves will normally double their weight during the winter on the Southern ryegrass.

When the heat starts to whammy the Southern pastures in May, their cattle start the trek north to their Birdsfoot trefoil and alfalfa pastures in Wisconsin. By combining the best of the two regions, the two Midwestern graziers are able to put 450 to 500 lbs.

of gain on a steer. They are sold in the early fall at 750 to 800 lbs. This high gain per animal tremendously lowers market risk and can produce an impressive profit per head.

Both graziers winter graze on a per pound of gain contract which is traditional on high quality ryegrass pastures. Smith winter grazes his 700 head of stocker cattle in Alabama and Ludlow puts his 2700 head out near Hattiesburg, Mississippi.

Contract grazing is big business in Pearl River County, Mississippi, and each winter some 60,000 to 75,000 calves are grazed in the county for Midwestern and western graziers. The gain rates are loosely tied to the price of corn. When we spoke they were being quoted for the upcoming winter in the 30 to 34 cents a pound range.

Ludlow is an ex-Colorado rancher who liked the high-return-per-acre economics of eastern grazing and moved to the rolling hills of west central Wisconsin in 1976. Dr. Larry Smith became Reid's Veterinarian and for several years they "partnered" on stocker cattle together.

Smith sold the cow-calf herd on his family's farm, switched to an all stocker seasonal operation like Reid's and retired from his vet practice to become a full time grazier. "We never could make any money with the beef cows because of their high wintering expense," he said. "A seasonal stocker operation is much better suited to the upper Midwest."

Today, Reid and Smith are just close friends but still enjoy comparing notes. The use of Southern winter pastures is not the only practice they have in common. Both use intensive grazing and both are big believers in Birdsfoot trefoil as a base for high gain potential summer pastures.

Birdsfoot Trefoil Tips

Birdsfoot trefoil is a non-bloating perennial legume. The problem most graziers have is getting it established. "Birdsfoot seems to grow where you don't want it, but won't grow at all where you want it," Smith said. The problem is competition. Birdsfoot just can't take it in the establishment year.

Using Roundup® on your pastures is the best way to establish Birdsfoot, Smith said, but he considers it too risky. "What happens if you don't get a seed catch and you've killed all your grass and you've got 700 hungry cattle on their way up from Alabama?"

His current practice for stand establishment and rejuvenation is to chisel plow his predominantly Orchardgrass pastures in the fall and then disk lightly in the spring before seeding the Birdsfoot. This has worked well for him but he admits Birdsfoot stands will frequently need rejuvenating.

Smith started out with large 40 acre radial paddocks that were popular in the early days of intensive grazing. He has subsequently squared up his paddocks as he found the radials caused excessive treading damage and killed out his treasured Birdsfoot trefoil. He has also reduced his paddock size to four acres.

He stocks these pastures at 2.5 to 2.75 calves per acre. The pastures are subdivided and the cattle shifted daily or every other day. He expects the Birdsfoot and Orchardgrass pastures to produce 300 lbs. of gain per steer during their graze in Wisconsin. In 1992, he had one herd of 400 heifers and one herd of 300 steers being rotated as two separate units.

He also grazes some predominantly alfalfa pastures. He has had his stand for seven years with no additions in P and K. **"The trick to lowering fertilizer costs is to not provide shade near the water source. All of your P and K will wind up under the shade tree if you let it,"** he said.

He found the best weight gain occurred if you timed your grazing to when the alfalfa was just starting to bud. He said his alfalfa produced an average daily gain of two pounds on steers and a gain per acre of 500 pounds.

Birdsfoot And Brome

Reid Ludlow said he likes a pasture mix of Birdsfoot, brome, white and red clovers. He stocks these pastures at a rate of two steers per acre until mid-August when half the cattle are sold and the remainder are shipped at varying times between late September and early November.

Both Ludlow and Smith emphasized the necessity of selling any Brahman blooded feeder cattle before September to avoid heavy discounts. Reid said he frequently "partners" with a cattle

103

feeder (sell some, have some custom fed) on his cattle to get a better price.

His 2700 head are divided into herds of 500 to 700 head and are rotated through 30 paddocks. Each paddock is shredded at least once each year to maintain quality and reduce weeds.

Ludlow said he likes to assemble his own cattle and check on them frequently during the winter. He said a nice side benefit of a seasonal stocker program was a tax deductible winter vacation.

Larry Smith said that as a native of the economically depressed Kickapoo region of Wisconsin he had hoped he could wake his neighbors up to the value and profit in pasture but it hasn't happened yet.

"We can take this steep land that won't produce 40 bushels of corn and consistently net $200 an acre in grass year after year," he said. "Why they can't, won't, and don't want to see that is sure a mystery to me."

Chapter 22

Managing for Quality Browse

Claudia Ball's ranch was one of the first to practice small pasture rotation in Texas. Her grandfather, Claude Hudspeth, Texas' "Cowboy Congressman," purchased the ranch in 1915 and had it subdivided into many small paddocks. Unfortunately, after his death, a subsequent ranch manager found all the fences bothersome and tore them out in 1941. The subsequent lack of grazing management allowed the ranch to deteriorate to a brush-covered pile of rocks with virtually no grass cover. With little to no grass to manage, Claudia has made the management of "quality browse" her goal in restoring the old family ranch.

Beef cattle are a minor part of her livestock mix, with sheep, angora goats, and commercial deer hunting her primary products.

"I have found that your better browse species need an even longer rest period than your grasses," Claudia explained. "You need to plan for at least a 90 day rest period if you are planning for quality browse production."

Claudia has found that Spanish goats (meat goats) are

easier to fence in with electric fence than hair goats. A three wire fence will hold a hair goat in flat country. A goat can tell if the electricity in the fence is on. A sheep can't, she said.

Goats are harder to train to rotational grazing than cattle or sheep. Claudia said that goats like to hide their kids while they graze and always want to go back to the previously occupied paddock. On her ranch, cowboys drive the hair goats to make them move.

Sheep should not be allowed to graze in river bottoms in areas subject to flash flooding. Cattle will get out of the way of the flood but sheep will drown. On Claudia's ranch, beef cattle are primarily used to graze a river bottom subject to such floods.

In areas heavily populated by deer, all electric fences should be kept on to prevent damage from deer. Since she started managing the ranch for quality browse production, her deer population has increased to one deer for every four acres! (This compares to one deer to 15 acres in the woods of Mississippi.)

Claudia has been replacing all of her windmills with solar-powered pumps made by A.Y. McDonald in Dubuque, Iowa. She says these pumps have no leathers or oil to replace and can pump as much as nine gallons a minute on bright days.

Chapter 23

Easy Does It...

I asked Alan Stallings to stand near his one wire, electrified fence for a picture and a hundred curious stocker cattle gathered along the fence to watch and make sure they were among the first in line in case of a paddock shift.

"It's hard to get used to just calling and having the cattle come to you," Stallings said, noticing the gathering crowd. "It's sure easier for one man to run more cattle today than it once was."

Stallings' family had always had beef cattle on their family farm that overlooks the Arkansas River near Morrilton in north central Arkansas, but they also had a feed mill and grain elevator where labor could be "borrowed" when it was time to work the cattle. Today, Stallings has sold the feed mill and elevator and "retired" to his ranch. The labor for the 100 momma cows and 200 stocker cattle he traditionally runs comes from himself and one part time hand.

For the last several years, Stallings has poured time, money, and sweat into fresh water reticulation and pasture subdivision. A cooperator with the Noble Foundation in Ardmore, Oklahoma, Stallings has built a 16 paddock cow-calf cell designed by Charles Griffith of the foundation. He also has a temporary 9 paddock polywire cell for his stocker cattle on winter annuals and clover.

The cow-calf cell is built on a steeply rolling bluff that overlooks the river, and the stocker cell is built on the river flood plain itself.

The cow-calf cell's grass is bermuda and fescue and the stocker cells have wheat, reseeding annual ryegrass, and Arrowleaf clover. Only the wheat is seeded each year. The naturally fertile bottomland requires no fertilizer and the clover provides the nitrogen, but ranching on a flood plain can have its moments. "In 1986, our whole stocker cell went under in a flood for two weeks. The flood killed out all the wheat, but the ryegrass came back like a charm."

The temporary, polywire fences allow his stocker pastures to be plowed and reseeded to wheat each fall. "I've tried planting fungus-free fescue in the bottom to keep from plowing but the ryegrass took it out."

Stallings had hoped that his cow-calf cell would allow him freedom from hay and feed, but it hasn't worked out that way yet. He can get to early February with no hay, but he has figured a four week hay feeding period would be needed to get him to spring pasture. In 1989 he planned to protein supplement his cows earlier as he felt he let their condition get too low in 1988.

"I seriously thought about going all stocker last year, but with the prices of calves what they are, I guess I'll stick with them awhile longer," he said at the time.

His philosophy is not to push his ranch for maximum per acre production but to work toward making the margins per pound of beef produced as wide as possible.

"It would be nice to say I produced a thousand pounds of beef per acre, but a man my age can't take that kind of financial risk. There's a lot of risk management in a very low input cost per pound produced," he said. "Easy does it, is what does all right by me."

Chapter 24

Heifer Grazing Gives Grazier A Mid-winter Vacation

Several years ago when Joseph Hicks' dairy barn burned down it looked like the end of the road for his life on the farm, but like many of life's tragedies it really was just a turning point. He discovered grass farming and found a very energetic city-born wife, Becky, who could move mountains with her enthusiasm for the country life. They decided not to rebuild the dairy barn but to try and create an enterprise that was suited to the lifestyle they wanted.

Joseph and Becky are big believers in the need for an annual vacation to keep one's sanity. They certainly understand dairymen who want to stop milking for at least a couple of months every year, but the Hicks have gone beyond that. They have **no** cows, dry or wet, to feed for six weeks each winter. "Everything, and I do mean everything, goes in December," Becky Hicks explained. "We shut our whole operation down for six weeks each winter, catch our breath, and start back up in February. We used to try and overwinter the tailenders but we put a pencil to it and it just wasn't worth it."

The Hicks buy in a couple of hundred 500 to 600 lb. Holstein heifers in February, grow them on pasture all summer and sell them as spring calving replacements in the fall and early winter. The heifers are bred natural service to Holstein bulls.

"These heifers are bred to fit a seasonal grass dairy. We would like to use Jersey bulls but the market so far insists upon Holsteins," Becky said.

Their February start requires about six weeks of silage feeding but they now buy this from a neighbor to avoid the need for machinery. Their sole piece of farm equipment is a John Deere six-wheel ATV.

One unique aspect of the farm is that their pasture is subdivided with a Harry Wier designed Technosystem. Wier is a leading New Zealand Holstein bull grazier. The Technosystem was designed to keep bull mobs small but still allow one grazier to handle several hundred bulls at a time.

The Technosystem features a practically gateless subdivision. The fence wire is connected to spring tensioners that allow the grazier to create a gate wherever she wants it just by stepping on the wire (with electrician boots).

These spring tensioners allow ATV's with special skids attached to run over the fences at will. It is quite unsettling for a newcomer to go zooming over paddock fences at 25 miles an hour but it greatly speeds up the productivity of the grazier.

The Hicks use an orchardgrass, perennial ryegrass and clover pasture as their base but also lease crop aftermath acres for fall and early winter grazing. They overseed their neighbor's oat stubble with five pounds per acre of Forage Star turnips in mid-August. The heifers graze the turnips, volunteer oats and adjacent corn stalks from the first of November until mid-December. This forage combination produces an average daily gain of three pounds a day. "You've got to look for such situations. I call it the fat of the land. There are so many opportunities out there for graziers today," she said.

Chapter 25

As Good As It Gets

You don't have to scratch most west Texas ranchers very deep to find the common fantasy of an existing dream ranch somewhere over the border in a remote valley. I hate to feed the fantasy but those fairy-tale ranchers do exist. I know. I've been there.

Hacienda Las Pilas, owned by Guillermo and Doris Osuna, occupies a remote valley in Coahuila, Mexico some 73 air miles from Del Rio, Texas. To reach the ranch takes 8 hours by road or 45 minutes by air from Del Rio. The ranch was originally owned by German interests but was confiscated by the Mexican government during World War II at the urging of the United States. Guillermo's father purchased the ranch from the government in 1947.

"This is absolutely the best grass year we've ever had," Guillermo smiled as we spiraled down to his landing strip in the green valley in 1987. The Gulf facing slopes of the surrounding mountains were thick with jungle-like growth and the valley floor swayed with healthy green clumps of bunchgrass.

He explained that the ranch normally received 18 to 20 inches of ran a year. In 1986 it received 50 inches and received 15 inches in the first six months of 1987. The excellent grazing conditions combined with 1987's high prices of calves on the export market had Osuna and other ranchers in northern Mexico saying little bad about the cattle business.

The Hacienda's main house is of modern design and is separated from the guest houses and swimming pool by a swinging bridge across a canyon, reminiscent of an Indiana Jones movie. A satellite dish and radio telephones help to break the isolation, but the education of the children is a problem for Osuna and the other ranchers.

His own children were educated at the ranch by tutors until they reached the sixth grade, then attended school in Del Rio where the Osunas maintain another residence. Airplanes are used by the Coahuila ranchers the way most U.S. ranchers use pickup trucks, but it's still a long way if you need a pack of cigarettes.

To emphasize the isolation, consider that Osuna lost 24 calves to mountain lions in 1986 and 7 by August 1987 when I visited. Bears frequently prey on colts and calves also, but Osuna takes these losses as part of the natural cycle. Deer and turkey are so thick and tame you could kill them with a rock, but Osuna prefers to watch rather than hunt them.

Limiting Factors To Land Ownership

The Mexican government limits the amount of land ownership allowed an individual to 500 animal units or around 120,000 acres in northern Mexico. To help with economy of scale and efficiency, Osuna has a management company that manages Las Pilas and eight other ranches. Las Pilas is stocked with registered Beefmasters and produces seedstock for use on his other eight ranches and neighboring ranches.

An early convert to HRM, Osuna operates Las Pilas as an 18-paddock cell. Water for the entire ranch is piped from a high mountain spring. In the three and a half years he has operated his cell, Osuna has seen the plant diversity multiply ten-fold. Johnson-grass, dallisgrass, Sideoats grama and Indiangrass now grow where they were absent before. Also obvious is the dying off of prickly pear and Flameleaf Sumac brush.

The valley floor is nearly at the four thousand-foot level. The summers are quite cool and the winters can be cold with an occasional snow. He has shifted from an early spring calving season to a late spring season to bring his breeding period more in tune with the natural grass cycle. No hay or supplemental feed is given to the cows.

Osuna said that ranchers in northern Mexico traditionally either sell their steer calves to the U.S. or send them to Monterey, depending upon the prices. It costs some $8 to $10 cwt. in fees and taxes to "cross" a calf to the U.S. and this traditionally keeps the two markets in equilibrium or the Mexican prices slightly at a premium. No females or breeding stock can be exported and the Mexican government issues export permits by state to insure that the U.S. does not bid away cattle needed for domestic consumption.

The extremely high prices of calves in the U.S. pulled a larger percentage of Mexican calves to the border in 1987 and into the export market. The U.S. government puts no restrictions on Mexican cattle exports, but Osuna worried that the very high number of export calves that year could spark some kind of American retaliation.

"The importation of a 400 lbs. Mexican calf that will be grazed and fed in the U.S. creates a lot more wealth for U.S. ranchers than the importation of a ton of boxed beef from New Zealand or Australia," he said in defense of the Mexican exports.

Osuna's love of nature has extended to his fencing and cell design. **He has taken the bottom wire off of all his barbed-wire fences to allow the free movement of game and keeps the bottom wire of his electric fences high also. He said he will build no more rigid, spoked-wheel type of grazing cells, but will use a more natural design that follows the natural contours of the land.**

"I want my cattle to flow naturally through the cell with no stress," he said.

Osuna knows that one day the rains will stop and the price of cattle will fall again, but in the summer of 1987, ranching in northern Mexico is about as good as it gets anywhere in the world.

Chapter 26

Sheep Dairying Can Produce
A Quality Life from Small Acreages

Tired of $11 cwt. milk and terrified $8 cwt. is on its way? How would you like to be selling your milk for $60 cwt.?

Well you can. All you have to do is switch species. To sheep!

Diane Kaufmann of Chippewa Falls, Wisconsin, is one of a growing number of graziers in the upper Midwest who have started sheep dairying recently. She was already into selling lambs, wool and hides and saw cheese milk as just another product for her ewes to produce.

Well, it was not really that cold a calculation. Diane surprised me by saying that a sheep dairy has always been her "life's dream" since she was a little girl.

Diane's ewes lamb in April and May and the lambs are allowed to have all of the milk the ewes produce for 30 days. After that the lambs are weaned and the ewes are milked twice a day for an additional 120 days. She said the ewes will average just over a pound of milk per day for the 120 days. Unlike France, sheep in

North America have not been selected for milk production, and production from one ewe to the other can be highly variable with some giving as much as a pound and a half. "Sheep dairying is ideal for the grass-based producer," she said. "A 150 day lactation almost perfectly matches our growing season in northern Wisconsin."

She has found sheep milk normally fluctuates between 40 to 80 cents per pound. Figuring an average price of 60 cents a pound for the milk translates into an additional $70 income per ewe per year.

The milk is sold to a cheese manufacturer in Minnesota. Sheep milk is so high in solids it can be frozen without damaging it for cheese making. Her milk is frozen in sanitary plastic bags and shipped to the cheese plant by UPS!

"The United States imports 43 million pounds of sheep milk cheese each year," Diane pointed out. She said the most popular in the United States was Roquefort.

By renovating one corner of an existing barn, her husband, Greg, was able to build an approved sheep dairy parlor for around $6000 including the freezer for the milk. The dairy equipment, which looks like a miniature cow milking machine is primarily imported from France and is readily available. One unique feature Diane loves is a self-locking head stanchion that allows Diane to milk the sheep alone.

The sheep quickly adjust to the milking routine, she said.

Diane started out milking 22 ewes in 1992, moving up to 40 in 1993. Her goal is 100 ewes, which she figures is about the limit of her family's 22 acre farm. With the wool and hides paying most the out-of-pocket cost of the ewes, the lambs and milk are almost all free and clear.

Five ewes per acre would net around $750 an acre, which is almost comparable to a seasonal grass cow dairy, but can be built for a fraction of the cost. "And one only has to milk for four months rather than nine or ten," Diane emphasized.

She markets her ewes' lambs direct to local customers and has a thriving pastured poultry business as well. In 1992 she raised 900 chickens on pasture and planned to double that in 1993. "I've doubled my poultry production every year. Selling pastured chickens are absolutely no problem. People stand in line for them. The lambs, unfortunately, are more difficult. You have to create a market for them," she said.

In 1993 she was selling lambs for 85 cents a pound live-weight and the chickens for $1.25 a pound up to five pounds and $1.75 a pound for those over five pounds. She closely follows Joel Salatin's recommendations on pastured poultry and uses his design of movable grazing cages with the chickens following her sheep through the paddocks.

Diane hopes high return per acre farming like sheep dairying and pastured poultry will allow rural women an alternative to having to leave their children during the day for a job in town.

Chapter 27

Mississippi Ranch
Takes on Change in Big Bites

Bob Meucci is physically a very big man who is willing to take on change in equally big bites. A successful pool cue manufacturer, Meucci is attacking convention in the conservative American beef business with an almost Kamikaze-like fervor.

His 1000 acre ranch in the sharply rolling hills of Northwestern Mississippi near Memphis is a showplace of the very latest intensive grazing technology. His most prized new forage is the cotton planters worst nemesis--Johnsongrass.

The cattle genetics he uses are a mixture of extremely rare breeds that would rock even the most hardened traditional cattle buyer on his heels and cause traditional seedstock breeders to blanche.

Thumbing his nose at the necessity of fat marbling for tender "quality" beef, he bypasses the traditional feedlot phase and direct markets his mostly grassfed meat to consumers at a considerable premium.

And whereas, most graziers would consider $600 to $700 a head an excellent price for a grass steer, Meucci grosses over $4000 a head!

Dizzy yet? Well, hang on!

Ruth Meucci, Bob's wife, said they started cattle ranching 16 years ago with a small 10 acre ranch in the Memphis suburbs. This mini-ranch was sold and the current ranch in remote Tate County, Mississippi was purchased. From an original 600 acres, it has grown to over 1000, but only around 500 acres are subdivided and used for grazing. These 500 acres produce the total feed requirements in grass and hay for around 570 head of cows, calves and stockers.

As very health conscious people, the Meuccis became interested in ways to produce tender beef with little to no fat. After many years of experimenting with various breeds and production methods, they believe they have finally identified the secret of tender meat and it is not fat. "Meat tenderness is a combination of rapid average daily gain and the lack of exercise," Bob said.

A New Breed

An early convert to intensive grazing, Bob said it allows him to replicate the sedentary environment of a feedlot but to do it on grass. His cattle spend their whole life on grass and are "finished" on pasture with a 60 to 80 day supplemental feed of high protein dairy cow feed rather than the traditional high starch grain ration.

"Protein grows muscle, grain grows fat. We found out we couldn't keep the fat off with grain on grass feeding even when we fed them only 60 to 80 days," he said.

The Purina Mills 18% commercial dairy feed produces an average daily gain on pasture of over three pounds per day at a cost of only 36 cents a pound of gain compared to over 50 cents in a commercial feedlot. The feed is fed on the ground under the electric fence to prevent the cattle from stepping or defecating on it. The cattle are rotated to fresh pasture daily just like growing stocker cattle.

All of Bob's male cattle are grown out and slaughtered as intact bulls at 1100 pounds at 15 months of age. By slaughtering them at this young age fighting and riding were not a problem. "We have found no difference at all between bulls and steers if they are slaughtered at 15 months of age. But we have found that cattle younger than 15 months are not as tender, so we don't want to push them any faster," Bob said.

The cow herd is divided into spring and fall calving herds to provide a year round supply of calves. Despite his need to spread his cattle out for slaughter purposes, he still aims for a very tight four to six week calving season to save labor.

The Meucci's are developing their own unique breed of cattle, which they call "American Blue." This breed is 34% bison (buffalo), 50% Belgian Blue with the remainder Angus and Charolais. They are basically a Belgian Blue, Beefalo cross.

Bob originally became interested in Beefalo for their muscling and this was also what led him to Belgian Blue.

The Belgian Blue is a double-muscled European breed genetically designed to produce a high percentage of high value steak cuts. Unfortunately, the double-muscling requires calving by Caesarean section and this limits its usefulness in the United States.

European-Style Meat

The Meuccis have concentrated on producing a low birth weight but genetically highly muscled calf that can be naturally calved.

By using European meat cutting methods that involve the careful disassembly of muscle groups, a Belgian Blue carcass can yield 60% of its carcass as high value steaks versus only 16% for traditional beef breeds.

Based on July 1, 1992 retail meat prices, Bob said a traditional USDA Choice steer would sell for $1345.50 at retail versus $3122.58 for a European cut Belgian Blue.

Thanks to a premium price as a result of his guarantee of 95% lean, no antibiotics and no steroids, in 1992, his "American Blue" sold for $4398 a head at retail--a $1200 per head premium over a grain finished Belgian Blue. Not a bad incentive for a pasture program.

"With a 95% lean carcass there is no trim to throw away. Everything is edible meat," he said. Special cooking instructions are included with the meat so people will not accidentally overcook the extremely lean meat.

The carcasses are allowed to hang in a plastic bag for six days following slaughter and slowly cool. This prevents the meat from chilling too fast, drying out and becoming tough. The meat is not cut until the ninth day. The meat is sold frozen and is shipped as far away as Florida in insulated boxes.

Of course, Bob is the first to admit that a producer has to direct market his beef to benefit from his particular genetic package. In normal market channels, packers get to sell fat to retailers for the same price as beef and therefore have no incentive to pay a premium price for high yield.

Meucci said that anyone with a little imagination can direct market lean beef. His primary problem is supplying the market, not finding a market.

With the market demanding more lean meat than his current cowherd can produce, Meucci has contracted with dairymen to buy back over 700 calves produced from his "American Blue" semen. He is currently paying $225 for a day old calf.

"That may sound like a lot, but these calves grow so fast, that we've totally recouped our purchase cost by the time they are six weeks of age," Meucci said.

In fact, the contracting with dairymen for calves is working so well, he is seriously considering discontinuing his own cow herd. With his genetics stabilized, he can rely on frozen embryos and semen and liquidate his herd of cows. He said beef cows were labor intensive and produced a very poor return per acre of grass compared to the dairy stocker calves.

"Its hard to justify letting a cow, that will at most produce one calf a year, eat the grass that could grow out four stocker calves," he said.

He said that by liquidating the cow herd he would have enough grass to grow out 2000 to 2400 slaughter bulls a year.

The Grass Program

Meucci's base pasture is a mixture of bermuda, dallis, crabgrass, white and red clover. He is drilling in Johnsongrass as an additional warm season component and is very pleased with it.

For wintergrazing, he drills in Crimson clover and Triticale-- a genetic cross of rye and wheat that he said will grow at temperatures barely above freezing. He has had problems getting the correct blend of Triticale varieties for the last two years but that in years past when he had the correct blend he has had knee deep Triticale in January.

His base farm is divided into 100 paddocks and all fences on the farm are electric. He is installing buried stockwater and irrigation pipes in his paddocks. He said a lot of money can be saved if both of these water lines are put in at the same time.

Water Water Everywhere

The irrigated paddocks are long (crossfenced with temporary fence to form square paddocks) and 260 feet wide. This allows the irrigation gun of a reel-type irrigator to traverse the center of the paddock and completely cover the entire paddock.

An electrically powered well pumps into a surface pond that serves as a reservoir for both natural runoff from the paddocks and the pumped groundwater. A diesel pump then pumps the pond water to two portable irrigation units.

The irrigation line cuts across the end of the paddocks and has a connector for the irrigator. The irrigator will irrigate the left hand long paddock, be spun around and irrigate the right hand long paddock and then move to the next paddock.

The use of portable irrigators keeps the hardware cost low in high rainfall climates where irrigation is largely an insurance policy against occasional drought.

Meucci said the primary cost in irrigation is in drilling the well. He said the well alone cost $20,000 and the pumps, irrigation guns and underground piping cost another $30,000. He said the total cost of a paddock that is irrigated, plumbed for stockwater and subdivided with two wire permanent electric fence was around $400 an acre. This includes the cost of the well.

Stockwater and permanent electric fencing, without irrigation, costs around $60 an acre. He figures his labor at 11 cents a foot for stockwater and 12 cents a foot for electric fence. The stockwater is buried under every other paddock fence and has snap-on water line couplers every 160 feet. The use of portable water troughs on a 100 foot garden hose allows one water connection to water four separate paddocks.

With the reel-type irrigators, the gun carriage is pulled out to the end of its hose with a tractor and the reel then winds it back in like a fishing reel winds in its line and cork.

At the Meucci ranch the irrigator guns must traverse steep hills that prevent the tractor operator dragging the gun carriages from seeing the end of the paddock, and sighting poles must be placed on top of the hills. These sighting poles have become great places for Ruth to hang Purple Martin gourds for natural mosquito control. She has also planted fruit trees along the vehicle access roads so that space is not wasted as well.

Excess pasture is captured as hay in large round bales. The bales are stored at the end of the paddock where they were made so they will be sure to be fed back in the same paddock and recycle the nutrients. One temporary polywire prevents the cattle from getting into the hay until it is needed.

Bob has recently ordered a custom built tractor with high flotation tires that will allow him to feed hay in very wet weather with no tire track damage to his grass.

All pastures are clipped once during the summer for weed control and to maintain grass quality.

In keeping with the biblical admonition to "make it well with thy fields before you build your house" the Mueccis have put their money into grass and cattle and live in a modest double-wide mobile home. One extravagance they allow themselves is a money-losing registered horse operation.

"Ruth and I have agreed that we would rather reinvest our profits than pay taxes on them. The ranch is almost completely developed, if fact, we could stop now and it could easily pay itself out," he said.

Bob figures by the time he has finished fully developing his 1000 acres and putting in a small on-farm slaughter plant he will have nearly two million dollars invested in the farm but says it will still be an excellent investment.

"At 2000 slaughter animals a year we can sell the meat **wholesale** and still clear an easy $1.6 million a year. That includes the cost of the grass, the feed, the labor, the houses, **and** the registered horse operation," he said.

Then thinking a moment, he added, "Of course, I think maybe we should start looking at a long range goal of 10,000 animals a year. What do you think Ruth?"

Bob Meucci. Big man. Big ideas.

Chapter 28

Parker Ranch Mixes Intensive And Extensive Grazing

The Parker Ranch at Kamuela on the big island of Hawaii proudly celebrates its 148-year-old (in 1988) "Paniolo" (cowboy) heritage with ranch rodeos and horse races. There is an excellent museum complete with a film that illustrates a day in the life of a Parker Ranch "Paniolo."

Tours are available of the blacksmith shop, ranch saddle shop, stables, and corrals and are popular with the thousands of tourists that visit the ranch each year.

Hidden in a high rainfall valley near Kamuela, is another Parker Ranch. Here the Paniolos are called cell managers and they ride motor bikes and ATC's and measure rainfall and grass growth and speak the strange tongues of paddocks, residual dry matters, flogging, and pugging.

This other Parker Ranch is personified by Jenny DeSilva. Jenny is 21 and a recent graduate in animal science from the University of Hawaii. She admitted that the challenge of intensive grazing is the only reason she had not followed most of her

123

classmates to the more exciting economy of the mainland. As the first woman cell manager she feels the pressure to be just as good as, if not better than her fellow male cell managers. There are seven cell managers on Parker Ranch. These seven report to their own management division at corporate headquarters. Cell managers are hired for a 7 day week. No one tells them when to go to work or when to quit.

Cell Manager's Jobs

Jenny is responsible for 38 paddocks but she also has to keep her eye on her fellow cell managers' paddocks so that weekends and days off can be covered. She records rainfall, grass growth, and her management decisions daily. These are sent monthly to Livestock Manager, Robbie Hind, for review.

Each month, cell managers are eligible for performance bonuses based upon how well their cattle are gaining.

Jenny's job is to grow out replacement heifers in her cells. The heifers are brought in from the extensive ranges at weaning. Jenny said the cattle are like wild animals when they are first brought in, but she has them tamed down and taught the paddock shifting routine within 28 days.

Fencing crews are building new paddocks every day and there are now 4200 acres under intensive grazing in the valley. Like most Hawaiian ranches, Parker started with centered hub cells, but are now shifting to New Zealand blocks despite their higher costs of construction. Jenny said the centered hub cells cost $30 an acre to build and the New Zealand blocks $50 an acre due to higher water reticulation costs.

The New Zealand blocks are superior in the high rainfall valley in that they result in less nutrient transfer, pugging (bogging) and offer more flexibility in operation (leader-follower systems).

New cells have to go through a period of grass conditioning before they get up to speed, Jenny said. The Kenyan Kikuyu forms a dense mat of dead leaves and stems that must be removed with close grazing before the grass becomes stocker quality. Ideally a cell should be "conditioned" with dry cows prior to becoming a stocker or replacement cell.

The Parker Ranch has no haying or silage equipment so all conditioning has to be done with animals or fire.

David Ramos, the business manager of Parker Ranch, said he hopes that intensive grazing will allow running the ranch's cattle production entirely on its deeded acreage and give up its leased lands. This will shrink the ranch from 225,000 acres to 180,000 acres and will probably cost the ranch its "biggest under a sole ownership" claim.

Richard Smart, the ranch owner, now lives in Honolulu for most of the year and is not active in day to day ranch management. The financial affairs of the ranch are overseen by a Board of Directors. Ramos said the Board had directed him to put most of the ranch's new investment into intensive grazing and tourism development.

In 1988 the ranch had only one cattle employee per 1500 head of cattle, which is considered a good ratio on the mainland. Intensive grazing actually increased the number of ranch employees because it required a different kind of employee than cowboying. "We've found intensive grazing requires a younger, more self-motivated type of person," Ramos said.

Prior to the decision to go all out on intensive grazing, the ranch was feeling the growing shortage of labor in Hawaii and was actively recruiting mainland cowboys for the ranch. The mainlanders soon grew tired of ranching in paradise and invariably moved back. "We had to face the fact that there are very few young people today who want to be cowboys. This worries us because there is a lot of this ranch that is probably more economically ranched in an extensive manner," he said.

Climatic Extremes Of Parker Ranch

Earl Spence is a Parker Ranch agronomist who helped instigate intensive grazing on the ranch and ran its first experimental cells in the valley. Today, he is trying to develop revenue for the ranch at its other climatic extreme--the extensive desert and near-desert lands that lie in the rainfall shadow of towering Mauna Kea volcano.

"The problem with cattle ranching in these areas is the high cost of getting water to the cattle. We have to pipe water for miles

125

and miles to get it here. My current job is to try and develop income from animals who don't need water reticulation systems, and right now those animals are birds," he said.

The dryland areas support huge populations of quail, chukka, and pheasant that are marketed to mainland hunters on guided hunts for a pretty penny. Spence is developing forage crops for these flying grazers.

Another species Spence is interested in are meat goats. The lava fields and sides of the volcano all sport gorse and brush that would be excellent goat feed. Due to the island's growing Filipino population, the demand for goat meat is soaring and meat goats sell for between $100 and $140 a head.

The ranch at one time had a large sheep population, but they were done away with in the 1960's. Spence believes they have a place on the ranch today with herded flocks of ewes in the desert and the lambs in the intensive valley cells. Herds of feral sheep roam the side of the Mauna Kea and appear to do well with little water.

Kikuyu And Intensive Grazing

Spence pointed out clumps of orchardgrass and white clover that grow where the road crews have mowed the Kikuyu on the roadside. "That's what Hawaii was once and can be again with intensive grazing," Spence said.

Whenever the dense mat of Kikuyu is eaten back under intensive grazing, orchardgrass and white clover volunteer. The Kikuyu was thought to not spread by seed and was brought in as a forage test. It "escaped" and now dominates the higher rainfall areas of Hawaii and has pushed out the temperate grasses that grew before its introduction.

The high altitude desert in the saddle between Mauna Kea and Mauna Loa is now leased to the Marine Corps for tank warfare training and is no longer ranched.

The ranch owns a feedyard on the island of Oahu where it feeds its own and other ranches' cattle. The ranch cattle were all Hereford until quite recently, but are now crossed with Angus and Brangus.

Each year on the Fourth of July, the ranch has an employee

126

rodeo and horse race. The ranch also owns a shopping center and a lodge in Kamuela.

Two tours of the ranch are offered to tourists. One is an hour tour of the Puukalani Stables where visitors can walk through graphic displays that explain the details of beef cattle, horses, intensive grazing, and a typical day in the life of a ranch "paniolo." The tour also includes the blacksmith and saddle shop, the corrals, and the ranch homes.

A longer three hour tour actually takes visitors out to the intensive grazing cells in the valley and to the extensive desert areas still ranched in the traditional "paniolo" style.

The Parker Ranch seems to have successfully defused the potential conflict between intensive and extensive ranching cultures by admitting that both have their place on the ranch and both are worthy of respect.

Even at 148 years of age, the Parker Ranch is proving it can still learn new tricks.

Chapter 29

Beef and Potatoes
Profitable Mix in Northern Mexico

Despite my protestations that there was a perfectly good passenger train between Tampico and Monterey, my two American traveling companions talked me into flying.

We left Tampico in the midst of a violent thunderstorm in a pregnant-guppy-looking, twin-engine Fokker. After an hour and a half of being strapped into a plane that was tossed all over the skies and struck by lightening at least once, I had two enthusiastic converts to the merits of railroad travel.

The huge storm signaled that a major cold front was blowing through and the weather in the highlands of Northern Mexico was going to be far different from the tropical coast. By the time we reached Saltillo, the temperature was in the mid-40's and falling fast. Luckily, I had brought both a sweater and a coat as one of my traveling companions thought he was going to the tropics and brought neither. He was very thankful for the loan of my coat.

Dr. Fernando Cabazos, a leading large animal vet in Saltillo, had invited us to tour the irrigated dairy and stocker pasture

128

region in the area between Saltillo and near Torreon. A "few" of his clients who were interested in intensive grazing might join us, he told us at breakfast.

By mid-day, those "few" had grown to over 50 ranchers and our caravan of 15 suburban and pickups, plus a school bus from the local ag school, made quite an impression on the many tiny desert towns we drove through on our tour.

Torreon is a major dairy production area with some 60,000 dairy cows that primarily ships to Mexico City and Acapulco. In April 1992, the milk brought $15 cwt, a slight premium to Texas milk but considerably less than the Mexican Gulf Coast region.

Surprisingly, while only a couple of hundred miles from the Texas border, the area agriculture has more of a California feel. With an annual rainfall of only six to eight inches a year, irrigation and high value agriculture rule. Dairy, potatoes, onions and tomatoes are the primary enterprises.

The non-irrigated lands are extensively grazed by goats and hardy beef cows. Four-wing salt bush is a primary forage on the dry ranges.

Due to the elevation, the area has cool summers that allow irrigated cool season grasses to grow virtually year round. However, there is frequently a day-to-night temperature fluctuation of as much as 70 degrees F. Death losses on imported Wisconsin Holsteins average around 10%. Local dairymen said they find California dairy cattle much better adapted to the high, dry climate.

Unfortunately, high capital, US-style dairy technology rules Northern Mexico and has been exported with all of its attendant financial misery intact. Mexico is currently losing dairymen almost as fast as Wisconsin despite a chronic milk deficit and record high prices. I saw little direct grazing of lactating cows (green chop) but almost all replacement heifers and dry cows are grazed.

Beef And Potatoes

Recently, there has been a shift away from horticultural crops to stocker beef production rotation with potatoes. To prevent soil borne potato diseases, the potatoes are grown only one summer out of three. The land is kept in pasture the rest of the time. The potatoes are sold in Monterey to potato chip manufacturers for the local market.

Winter annuals grown in combination with alfalfa are a popular forage mix at the hotter, lower elevations. Cool season grass and legume mixes are used at the higher, cooler elevations.

Three new 25,000 head feedlots in the Monterey area have created a demand for a year round supply of feeder cattle. These are almost totally heifers as the tradition has been to export the higher value steers to the United States at weaning.

Beef heifers traditionally go on feed at around 650 lbs. and die at about 850 to 900 lbs. after 80 to 90 days on feed. It is against the law to feed corn to ruminants in Mexico, so milo and horticultural byproducts are used.

Graziers can contract to grow stocker cattle on a per pound basis or on a guaranteed buy-back program with the feedlots. With an average weaning weight of around 450 lbs., most of the heifers need 200 to 250 lbs. of grass gain put on them prior to going on feed.

Christmas Valley

Probably the most dramatic concentration of converted horticultural farms is in Valle de Navidad (Christmas Valley) in the state of Nuevo Leon. This valley is only 20 miles long and has the highest concentration of well-done intensive grazing I have found to date anywhere in North America.

Farmers in this valley are converting about 30 center pivots per year from horticulture to stocker cattle due to the higher profits per acre from grazing.

With an elevation of 6000 feet, the summers are cool enough to allow the use of perennial cool season grasses like bluegrass, fescue, perennial ryegrass, orchardgrass, and prairiegrass (rescue). However, these perennials are over seeded in late August or early September with cool season annuals like cereal rye, wheat and annual ryegrass. These annuals tremendously boost mid winter production and make year round grazing much more feasible.

These high production animals combined with intensive grazing and irrigation, allow Christmas Valley graziers to produce up to a ton of beef per acre and explains why beef is replacing tomatoes and onions.

A leading grazier in Christmas Valley is Fernando Elizondo of the 2470 acre Aguatoche Ranch. He said grass is not only more profitable than tomatoes and onions but is a soil healing crop as well. **"We have seen our soil's organic matter increase from 1.5 to 6% since we started a two-years-of-grass and one-year-of-potato rotation. That means we need to use less water. The grass also helps prevent sodium buildup in the soil (from the irrigation water) as well."**

Elizondo said electricity was used to run the irrigation pumps and was very expensive in his area. He has recently installed drip irrigation for the potatoes, but uses both side-roll and center pivot irrigators with the grass but prefers the center-pivot system.

He said once the underground piping needed for side-roll was figured in, the cost per acre was about equal to a center-pivot but the labor needed to operate it was much higher.

"We graze over 8000 heifers a year with only three cattle employees. The real profit in grass and cattle are the savings in labor and equipment compared to horticultural crops," he said.

Fencing Sudivisions

All interior subdivision is done with temporary fence to allow the land to be converted into potato production easily. With the center pivots, the pastoreador (pasture master) subdivides the circle with temporary one-wire fence into pie-shaped radials. These are further subdivided with a temporary cross fence to force the cattle to graze the front half of the paddock in the morning. They are then allowed the back half in the evening.

After shifting to the fresh paddock in the morning, the pastoreador takes down the fence from the previous day and rebuilds it in front of the occupied paddock. All stock water and minerals are at the pivot point in the center of the irrigator.

With the side-roll irrigators, the heifers are given a hectare of fresh grass each morning and the back fence is moved forward every fourth day. A side lane is used to allow access to stock water and minerals. The permanent pasture fence provides one side of the lane. One-strand of polywire is laid out in an L shape to provide both the back fence and the other side of the lane fence.

Water and nitrogen are always applied within two days of the cattle leaving a paddock. A seasonal average stocking rate of around 400 heifers per pivot (180 acres) is used, but this is raised or lowered to match the grass growth conditions.

"We graze around 8000 heifers a year," Elizondo explained. "This is roughly two 120 day turns of 3100 heifers and a slightly lower stocking rate turn in the winter. The heifers will gain slightly over two pounds per day while on our pastures.

"Our goal is to produce six kilos of gain per day per hectare averaged over 365 days. (4818 lbs. per hectare or over a ton of gain per acre!) We are currently achieving this goal," he said.

Elizondo's base pasture is an eclectic mix of grasses and legumes designed for high alkaline soils. This pasture salad includes Strawberry clover, Birdsfoot trefoil, perennial ryegrass, fescue, bluegrass, orchardgrass and brome. This particular mix is marketed as "St. Germaine's Mix" by Douglas-King Seeds of San Antonio.

To this he has added a new alfalfa-like legume called Sainfroin that is bloat-free, cold tolerant and specially bred for high pH desert soils. However, with all of these perennial grasses and legumes, he estimates that half of his forage production is still from the overseeded annuals--rye, wheat and annual ryegrass.

"Keep in mind that irrigated pasture is expensive pasture. You must use high production grasses, high stocking rates and high value producing animals to make the most of it."

Chapter 30

Polycultural Grazing in Arkansas

Ed Martsolf and I walked up to a mixed herd of sheep and pigs grazing together on a paddock of sweet potatoes and cowpeas at Heifer Project International near Perryville, Arkansas. I had never seen this particular combination of grazing animals and plants before and wanted a closer look.

However, before we got within a hundred feet of the sheep the long-eared Jenny that serves as their protector decided that Ed and I had gotten close enough and quickly gathered up her flock and hustled them to the far corner of the paddock. She then turned her rump toward us in typical donkey insolence.

The pigs on the other hand singing a chorus of oinks and squeals came running toward us as fast as their stubby legs would carry them. Like a bevy of your best beagles the pigs then proceeded to accompany us on a tour of their paddock by staying as close underfoot as possible.

"We call them our happy pigs," Ed explained. "The grazing pigs seem so much tamer and animated than our confined pigs."

The pig and sheep combination is not the only unusual mix at Heifer Project. Cows, sheep, goats, donkeys, pigs, chickens and ducks are all run as foraging animals in mixed herds at the ranch in the Fourche River Valley of central Arkansas.

Heifer Project International is a non-profit church supported (primarily Methodist and Presbyterian) mission to help feed poor people in the under-developed areas of the world through the donation of non-grain eating animals. The recipients of these animals have to agree to give a neighbor one of the female off-spring of the animal.

Heifer Project's History

"A cow not a cup" and "pass on the gift" are the ideological foundations of the Project. These foundations were set by Indiana farmer, Dan West, who started Heifer Project after serving as a volunteer relief worker during the Spanish Civil War in the late 1930's. West saw the food handouts were not only quickly exhausted but undermined the dignity of those recieving them.

Following World War II, the Heifer Project sent hundreds of boatloads of heifers and other farm animals to Europe. Today, the project works primarily in Latin America, the Carribean, Africa, Asia and the Southeastern United States.

A 1200 acre ranch west of Little Rock, Arkansas, is used both as staging area for live animal exports and as an income generator for the Project.

"We run the ranch to make a profit for the Project. While we now focus entirely on low-input farming methods as a matter of policy, this decision was largely forced on us by the hard times in agriculture a few years ago," Ed said. The primary dollar earner for the Project is a herd of several hundred registered Brangus cows. This herd is managed by Jim Combs and Cathy Pallatinus and produces seedstock quality bulls for both local sale and export.

The ranch is currently divided into 35 paddocks which are rotationally grazed. Bermudagrass, annual ryegrass and fescue provide a virtual year-round pasture program but a lot of hay (baled pasture clippings) is made to maintaing pasture quality.

A low-input commercial cowherd is being added both as a research project and as a way to more fully utilize the pastures.

Also a much more intensive pasture subdivision program for stocker cattle is underway. Ed would like to see the Project get into a seedstock program for pastured dairy cows as well in the near future. "We are researching high-quality plants that can be used for finishing slaughter cattle, pigs and lambs on pasture. We are working with the Kerr Center for Sustainable Agriculture on Eastern gamagrass, for example," Ed said.

Eastern gamagrass is an extremely high quality native grass that is genetically similar to the corn plant but is a perennial. Pigs, as well as cattle and sheep, apparently love it literally to death, unless it is rotationally grazed.

"Nature is growing a huge forage crop here in Arkansas that we are barely utilizing and yet Arkansas imports the hamburger meat we eat from overseas! I feel hamburger is a market the Southeast can fill at a profit," Ed said and indicated the Project would be interested in trying to help a ground beef producers co-op get started.

Easy Keeper Sheep

The Project also has the largest flock of Katahdin sheep in the country. The Katahdin sheep are a hair sheep bred for easy-keeping and a superior carcass. "Katahdin sheep are really just miniature beef cattle," Laura Callan, Director of Livestock explained. "There's no shearing, crutching, or tail docking with Katahdin as they are bred strictly for meat production."

The breed was started by Michael Piel in Maine in the 1950's when he began crossing Virgin Island hair sheep with Suffolks. These crosses were then crossed with a shedding breed from England known as the Wiltshire Horn to form the basis of the Katahdin breed.

While originating in a northern climate, the breed has shown itself to be exceptionally well suited to the Southern U.S. and the tropics. **The breed is both more heat and parasite tolerant than the wooled breeds and is a non-seasonal breeder. The Katahdin is more browse-oriented than wooled sheep and because it is not trying to divert energy into wool production can do well on lower quality pastures.**

"This is the ultimate easy-keeper breed. The Katahdin does

135

not require the intensive level of management some wool breeds require. All in all, we believe the Katahdin is the ideal sheep breed for the South," Laura said. Shepherd Mary Van Anrooy, said the sheep are bred to lamb twice in three years and the flock at Heifer Project lambs in February, May, and October. The lambs are weaned at 60 days and finish at a liveweight of between 100 to 120 lbs. The breeding ewes are maintained solely on pasture and are fed no grain.

On Guard

Female guard donkeys have been effective at preventing predation. However, the ultimate sheep guardian the Project has found is a six-foot tall castrated male llama. Dogs and coyotes are apparently terrified by both the look and the smell of the llama who aggressively chases them. Unfortunately, he was so aggressive protecting his flock that he would not let the shepherd approach the flock either. As a result the llama has lost his flock and is now used for night sentry duty around the ranch headquarters to keep dogs, coyotes and two-legged predators away.

Pastured pig production is a relatively new enterprise for

the Project, but Laura said they have already found they can cut the grain input in a finished hog by 50 percent through the use of high quality pasture and legumes.

Sheep and pigs together is another research project designed to both finish lambs and feeder pigs with no harvested grain while improving soil fertility. A six acre field has been subdivided into seven paddocks with electric fence and each paddock is planted to a different combination of plants to form a forage system. Hopefully, the mix chosen will be capable of providing a long grazing season suitable for finishing both hogs and sheep.

The forages used were: Sudex, feed corn (staggered planting dates), lespedeza and rape, collards and turnips followed by cowpeas, sweet potatoes and cowpeas planted in alternate rows, oats and turnips followed by buckwheat, and sweet corn and peanuts in three different arrangements. Results were incomplete when I visited but looked encouraging.

Another new grazing critter the Project has found is the "Khaki Campbell" duck. This foraging duck will lay more and larger eggs than the highest producing breed of chickens and do so solely from forage. The duck will forage during the day but will always return at dusk.

The addition of free-ranging New Hampshire Red hens around the barns and corrals have virtually eliminated flies while serving as a low-cost source of eggs and meat for the staff.

Chickens are also used as garden tillers. The chickens are cooped up inside a small A-frame enclosure in the garden and are fed kitchen table scraps. The chickens in turn till, weed and fertilize the small section of the garden under the coop.

After four to six weeks the whole coop is moved to a new location and vegetables are planted on the old site. After harvesting the crop, the portable coop can be returned to allow the chickens to eat the vegetable remnants and recycle them into manure for the next crop.

The Project has converted a typical contract chicken house that one sees everywhere in the Ozarks into a low-input "diversified" agricultural enterprise growing hens, broilers, rabbits and feeder lambs in the same building.

The ranch also serves as a research and extension "campus" for even more radical no-input ideas for the underdeveloped world. A learning trail winds through the headquarters area where Americans can learn how no-input agriculture can be practiced in the Third World.

Scrounge Farming

"It was hard for Americans to understand just how little people in Guatemala, for instance, had to work with so we have developed what we call our Guatemala farm. With some of our research volunteers we only provided nails and seed and they had to scrounge the rest, just like a farmer in the third world would have to do," Ed said.

This resulted in irrigation pipes being made from old Coke cans and fences from packing crates. A search for a local plant to serve as a roof thatch in the demonstration area discovered that locally abundant Johnsongrass is an ideal roof thatch material.

The Project recommends that each small farm should have three to five species of livestock, 10 to 12 annual crops and vegetables, and five to ten tree crops. Leguminous trees such as honey locust can produce both nitrogen for companion plants and high quality forage for both pigs and ruminants. A two acre farm the Project helped develop in Bangladesh is feeding 16 people with no industrial inputs.

Rabbits appear to be the best family meat-producing animal for the under-developed world. Two does can produce 150 lbs. of high-protein meat a year. **A primary drawback to rabbit production in the tropics is heat as rabbits will not breed in the heat. An underground rabbitry with a garden growing on top has been found as a way to beat the heat problem. The manure from the rabbit cages is then used to grow the garden on top.**

Systems whereby the outputs of one part becomes the input for another is the common property of all sustainable farming systems according to the Heifer Project.

In the Third World grain is too valuable as human food to be fed to livestock, so the Project's research into forage based livestock systems is useful both in North America and overseas.

Ironically because so much of the world looks to America for technological ideas, the adoption of "kinder and gentler" methods of farming in the U.S. could help stop the ecological destruction now occuring in many under-developed countries. However, this learning can be a two-way street as well.

Ed said the no-input Guatemala farm had also been useful in stirring up creative ideas among the American staff and workers as well. "You've got to realize that we're (the American staff) all the products of our high-input oriented agricultural university system. None of this comes naturally to us. We're having to struggle to rethink what we knew, or what we thought we knew, just like every other farmer in America."

Chapter 31

Way Down South...
In Hawaii

 $\mathbf{W}$ hile few folks at Kahuka Ranch near Naalehu on the southern tip of the big island of Hawaii go around whistling Dixie, the 12,000 acre ranch is as far South as you can ranch in the United States. The ranch is managed by Carl "Soot" Verdhoff for trustees of the S.M. Damon Estate. Soot is considered by many Hawaiian ranchers to be one of the best intensive graziers in a state where intensive grazing is a lot more advanced than on the mainland.

Like many ranchers in the U.S., intensive grazing has been "on trial" by ranch trustees. The original "trial" was a 266 acre cell near the ranch headquarters. The cell was designed by Earl Spence of Kamuela, Hawaii, and is a two-wire block design with water in each paddock and 26 subdivisions.

Wide "grazing alleys" are used to move the cattle and can substitute as separate paddocks if needed. The grass is primarily Kenyan Kikuyu.

In 1982, prior to subdivision, the 266 acres weaned 28 calves. Using 1988 prices, this returned the ranch 45 lbs. of beef per acre

140

and $28 in income based upon a $63 cwt. calf price.

In 1987, the 266 acres produced some 116,865 lbs. of weaned calf, or 239 lbs. per acre. At $63 cwt. this resulted in a gross production of $276 per acre or $75,625 on the 266 acres. The total cost of the cell was $25,296, including fencing and water reticulation--not a bad return on investment by anybody's standards.

The Hard Part

Soot, however, is the first to emphasize that what made Kahuku's first cell so successful was its small size and nearness to the ranch headquarters and his house. He recommends that **any rancher start with a learning unit that is small enough that it is easy to see what is happening in it, first.**

"Stringing electric fences and water lines are the easy part. Learning how to run it is what's hard," he said.

And the hardest part is not having any confidence about what is going to happen, he said. Kahuku's orientation until recently had been in seedstock cattle and the ranch's management was oriented toward individual animal performance rather than per acre production. This animal orientation had resulted in a management attitude that you gave the animal whatever it needed, but little to no thought about what the grass needed.

Intensive grazing and commercial cattle were a massive change to the ranch's way of thinking and did not come about painlessly. Soot suggested that a rancher new to intensive grazing needs to give himself and his fellow workers time to become acquainted with the new routine and also to watch the shift in pasture composition that occurs.

Running a new cell with cows is a good idea in Hawaii, he said, because the cows, rotation, and rest can condition the grass toward a quality composition. White clover, for example, began volunteering into the cell where it had never been seen before. However, continuing to run cow-calf exclusively in conditioned cells results in overly fat cows, he has found. Using stocker cattle as a "leader" segment before the cows would allow a more cost-effective use of quality pasture.

The ranch built two more 16 paddock cells across the highway from the original cell, but a severe drought caused them to use it as a single 32 paddock cell. When the drought ended, Soot continued to operate it as a 32, and in 1988 finished another new 32 paddock cell.

Soot has found that Birdsfoot trefoil will grow quite successfully on unfertilized soils with rotational grazing and is seeding this into the new cells for an added quality component.

The ranch in 1989 had 900 acres subdivided into 90 subdivisions. These are normally run as three separate cells, but two of the cells could be combined and run as one in a drought.

900 acres is not much on a 12,000 acre ranch, but Soot is convinced that a slow program that allows the building of confidence and competence is the best way to intensify grazing.

Chapter 32

Cow-calf Economics Shine
With Low-input Program

Clint Josey is hard to figure out, according to his neighbors. His pastures are grown up in tall clumps of Johnsongrass, switchgrass and bluestems. He doesn't mow to keep them nice and flat like they do. He doesn't plant sudangrass for hay. In fact, he doesn't even feed hay, or range cubes either.

The few neighbors who recognize native prairie grasses think he must have some special fertilizer to make them grow so thick, but Josey hasn't nitrated a pasture in six years. He has operated his ranch as a grazing cell for 5 years but doesn't have an electric fence in sight.

If his neighbors would just sit down and think about the things that Clint Josey **doesn't** do, they could get as excited about the cow-calf business as he is.

"All my life I heard that you couldn't winter cows on standing hay in the wet part of Texas, but that's not true once you go to native grasses," Josey said.

Josey's pride and joy is his Coastal bermudagrass bottom

that has volunteered to thick stands of switchgrass providing most of his winter feed. A native ryegrass of Mississippi called Marshall, and Arrowleaf clover provide a naturally reseeding protein supplement for his cows in the winter.

"I've found that non-nitrated Coastal bermuda is an entirely different grass than nitrated Coastal," he said. "If you stop the nitrogen, Coastal becomes not only easier to manage but also allows a much more diverse plant complex to compete successfully with it."

The Coastal bermuda/switchgrass bottom is the centerpiece for Josey's ranch, which also includes very dry, thin soils on the hills, rising some 200 feet above the river bottom. The ranch is now operated as a 27-paddock grazing cell but was originally subdivided for use in a registered cattle operation built around single sire herds that needed a separate pasture for each herd. All Josey had to do to go to controlled grazing was combine his cattle into one big herd using the existing subdivision.

Leader-Follower

Since each pasture subdivision has its own water source, operating the cell in a leader-follower fashion is easily accomplished. For example, when the calves are weaned in the fall they go and graze in front of the cowherd through the winter and the bulls follow the cows. Once the calves drop in the spring, the cows go to the front, the yearlings go second and the bulls third.

By maintaining as few herds of cattle as possible and using the leader-follower principal, Josey is able to match grass quality to the class of animal with very little hassle.

The good soils of the bottom are the key monitor point for Josey's rotation regimen. When the bottom is ready to graze, the cattle are brought back to it regardless of where they are in the rotation. Some of the dry hilltops are only grazed twice in a year and are used primarily as a winter feed source.

"Most people recommend a 20-day rotation on Coastal, but they are talking about nitrated Coastal," Josey said. "**Once you cut the fertilizer you need to go to a longer rotation, and when you do that you start bringing back your native vegetation and Johnsongrass.**"

144

Josey showed me how the Johnsongrass and switchgrass were gradually climbing up the side of the hills. On top of the hills the short native grasses were starting to shift to a tall-grass prairie complex.

"I am convinced that with a long, rest-rotation regimen we can convert all of the wet part of Texas to a tall-grass prairie every bit as good as that in eastern Kansas and Oklahoma," he said.

The ranch in 1987 ran one animal unit for every six acres in the summer, with an increase in winter to one animal unit per four acres by bringing in stocker heifers. This compares to an average stocking rate of one animal unit per 14 to 15 acres, which is considered normal in the Forth Worth area. All calves are grown into yearlings before being sold in the spring.

"If you can stop feeding hay and feed, the economics of a cow-calf operation change dramatically. I mean, it's just got to make money."

Josey warns that in making this shift to a high-succession pasture in the East means having to endure the ridicule of your neighbors and having to fight your own urges to mow and spray weeds.

"Around here," he said, "Johnsongrass and native grasses are considered weeds and letting your pasture grow up in them is called 'white-trash farming.' You've just got to close your ears and spend your time counting the money white-trash farming is bringing you!"

Chapter 33

Spring Lambing Ideal
For Low-cost Pasture Program

Peter Woods of Blanchardville, Wisconsin, has read all the literature on multiple lambing seasons and finds it all too complicated to be seriously considered. What works for him is a simple once a year spring lambing program that maximizes the use of his pasture and minimizes the use of his checkbook.

Wood figured out it cost him two cents a day to graze a ewe, 10 cents a day to feed it as hay and over 13 cents a day to feed hay and a small amount of grain. This little cost analysis quickly indicated the key to a profitable sheep operation was to maximize grazing days and minimize feeding days. "53% of the cost of producing a lamb is feed costs. If you want to control costs, that's where you start," he said.

Thanks to spring lambing, intensive grazing and the seasonal use of woodland as pasture, Woods has cut his hay feeding time for his 250 ewes to no more than two months of the year and his grain feeding to only six weeks.

He said an April lambing season perfectly meshed with two

periods of high quality cool season pasture. The lambs are born and grown on the high quality spring pasture and the ewes are bred back on the high quality fall pasture.

Another benefit of spring lambing is that the highest animal numbers correspond with the greatest availability of pasture. "At lambing we suddenly go from 250 animals to 900 animals which meshes perfectly with the highest grass growth rate of the year. It's almost as if nature intended it that way," he said with a wink.

He said **spring lambing was particularly important if breeds that produce a lot of twins and triplets are used. "If a ewe has to produce enough milk for two to three lambs she has got to have a lot of high quality pasture."**

Woods uses Finn-Rambouillet cross ewes, which are high in multiple births. These F-1 ewes are in demand as replacements and Woods sells most of his ewe lambs into this premium market.

Wisconsin's Grazing Year

Woods' ewes lamb in April. The lambs are weaned in June at around 30 lbs. and are then grazed ahead of the ewes in the pasture rotation. Summer pasture quality is kept high through the use of Birdsfoot Trefoil. He said he had found the trick in maintaining a good stand of Trefoil was to always allow it to flower and set seed in the establishment year and build a good seed reserve in the soil.

The lambs are allowed to cream the summer pastures and the ewes are allowed to lose weight. The male lambs are sold as feeders whenever the summer pasture starts to get short.

On September first, all the sheep are moved off pasture and are grazed in the woods with supplemental hay until October 20th. This allows the pasture to stockpile feed for wintergrazing.

The ewes are then bred on this high quality stockpiled pasture in November. The breeding season is 21 days long.

On December first, the ewes are bunched into a tight flock and the stockpiled pasture is rationed out. Some years hay feeding does not begin until March 11th but in other years it begins in early January depending upon how deep the snow cover is. Woods has found the ewes actually prefer to eat snow than drink ice-cold water in the winter and he doesn't worry about fresh water as long as there is snow.

Six weeks before lambing, Woods begins to feed the ewes 1/2 pound of grain a day. Grain feeding is continued for ewes with triplets or quadruplets until weaning.

Replacement ewes receive one pound of grain per day through the winter. Peter found it was important to always allow a mature ewe to graze with the replacements as a "role model" and he grazes four ewes with the flock of 200 replacement lambs.

Peter said his program gave him four income sources. His primary one is ewe replacements, and in 1992 he had a two year waiting list for his ewes. He also sells a few ram lambs as breeding stock but most are sold as commercial feeders. And of course, he has his wool income.

He said he could possibly increase income by finishing his wethers himself but this would require him to buy more grain. "I guess there's a little bit of laziness in that decision as well," he said. "We're in the sheep business primarily for a quality of life that what we're doing now has so far been able to provide."

Chapter 34

Battening Down the Hatches

Hopefully the widely predicted "debt depression" will never come, but down in Texas, a ranch management group is busy battening down the hatches on the properties they manage for trust owners "just in case."

Headed by Dallas businessman, Bunker Sands, and former seedstock marketer, Ronnie Farrington, CHTE is rapidly converting the properties under their management to low-input grass-based operations using the guidelines of Holistic Resource Management. CHTE currently manages four trust properties in Texas as well as others in Colorado and Hawaii.

On the Texas properties, all grain farming has ceased as has the planting of winter annuals and nitrogen fertilization. This has allowed the sale of several hundred thousand dollars worth of equipment and has financed the stocking of the ranches with cattle. Pasture subdivision and rotation has replaced nitrogen fertilizer and hay on the two East Texas properties under CHTE management that I visited.

The property that has been under HRM style management the longest is the 10,170 acre Ennis Ranch located in the Trinity River bottoms southeast of Dallas. This ranch has been subdivided into 235 paddocks and today runs three times the number of cows it did a few years earlier when hay and fertilizer were still being used. Another property, a rowcrop farm near Seagoville, was also taken over by CHTE to be converted to holistic style management as well.

Stockers Complement Brood Cow Operation

Kenneth Braddock manages the Ennis ranch and his brother, Richard, manages the Seagoville property. The Seagoville property will be developed as a stocker growing operation to complement the brood cow orientation at Ennis. Both properties use Johnsongrass as their base forage with volunteer annual ryegrass making up the winter feed supply.

As of 1988 the only two agronomic inputs being made on the ranch were the growing of switchgrass for seed increase and the planting of a new Mississippi-developed clover called Bigbee Berseem that has done extremely well in the prone-to-submersion bottomlands.

"We really don't know where succession will take us in East Texas," explained Bunker Sands. "We feel we will eventually develop a tall grass prairie complex, but we aren't certain. Johnson-grass and volunteer ryegrass serve our immediate needs well."

The Seagoville property has established Yucchi Arrowleaf by feeding it in with the cattle's minerals and by throwing a coffee can of seed in front of each paddock gate before a shift. Another trick, Kenneth and Richard use is to never plant more than 15 acres of new grass or clover in one place. If the plant is well-adapted to the area, the cattle will move it in their manure and naturally spread it at no cost.

"We put it out there and if it grows, it grows. If it needs a fertilizer truck to grow, it'll die on this ranch," Kenneth said.

Kenneth and Richard are closely monitoring some of their neighbors' success with fungus-free fescue in the river bottoms and may add this as a test.

The 235 paddocks on the Ennis property are run as three

500 cow units except during droughts when the herds are combined to allow an extra long rest period to the grass. The average size paddock at Ennis is 34 acres.

"500 pairs on 34 acres of blackland river bottom sounds much worse than it has turned out," Kenneth Braddock explained. "I was the biggest skeptic about controlled grazing on blackland there was. I swore it couldn't be done here, but it has worked out well. We now will actually slow our cattle in wet weather to purposely churn a paddock ankle deep as we have found it really stimulates the Johnsongrass."

Richard said that converting crop land to HRM style management was slower than pasture land. "When you are starting out with farmed land, you are starting at less than zero. Our production per acre fell in half at Seagoville when we quit plowing and fertilizing. We have not got that back up to 75% of its former output with no agronomic input other than pasture subdivision and grazing management."

Penny Pinching Pays

The Ennis property was originally fenced with a two wire ground/hot system similar to that being used in West Texas, but the second wire was found unnecessary in the wetlands of East Texas. When the Seagoville property became available, it was fenced by stripping the ground wire off the Ennis fence.

Such tight-fisted, penny pinching is a trademark of CHTE's management style. "The first thing we did was to sell $200,000 of equipment off this ranch. We got rid of all the junk and replaced it with a few pieces of new equipment that would need a minimum of maintenance," Kenneth said.

Unable to get a bid on an old combine, the unit was rigged for forage seed harvesting and now custom combines clover seeds on a share basis with local ranchers. Pecans that formerly went unharvested became a cash crop at Ennis. "We were able to finance our cowherd expansion totally out of equipment and pecan sales," Kenneth said.

Sands first became interested in HRM on a visit to Hawaii where he saw his first intensive grazing and electric fence. "When I saw a drive-through electric fence, I was totally sold, because I hate opening gates."

Sands attended classes in HRM in Albuquerque and paid for all of his employees to go through the classes. He talked Ronnie Farrington, a retired former seedstock sales manager, into helping him convert all of the trust properties under their management into HRM managed properties. Farrington also heads up an investment group that looks for short-term opportunities in cattle, primarily stocker cattle. "The best money is buying calves as small as possible and growing them as big as possible on grass," he said.

CHTE has found it much cheaper to lease wheat pasture from others than to grow it themselves when agronomic and machinery costs and the loss of liquidity from the investments is considered. Cattle are liquid investments. Tractors and land aren't. Farrington believes that maintaining maximum liquidity is the safest option in these financially turbulent times.

"I had 8000 cattle on feed when the market broke in 1973, so I have had a lot of personal experience in how fast things can change," he said.

In response to the 1988 high prices paid for bred cows, CHTE began selling all their cows over five years of age. In 1987 they sold all of their steers at weaning off the Ennis ranch and turned around and bought light-weight Mexican steers for $10 cwt. less.

"Liquidity allows you to capitalize on opportunity," Farrington said.

CHTE is experimenting with irrigated pasture on their Colorado property in hope of shutting down crop operations there. A southwest Texas ranch is being managed primarily for game with minimal paddock subdivision and intensification, and a southeast Texas property is being managed primarily for pine timber production and hunting.

CHTE now has all of its Texas properties out of debt and in the black. They are like the rest of us, they hope the storm will never come, but if it does, they're ready.

Chapter 35

Snow Grazing, Canadian Style

It comes as a surprise to most American graziers to discover that a portion of the province of Alberta has about the same winter snow cover as the Texas panhandle or northern New Mexico or Arizona. This lack of snow is due to the hot winds off the Rocky Mountains known as the Chinook.

As cold fronts race over the Rocky Mountains and start downward toward the Canadian prairie, they compress and heat the air in front of them. Known as "snow-eaters" these hot winds are why Canada's beef cattle industry is concentrated in Alberta.

Attracted by the relatively snowless winters, graziers from the States and Eastern Canada built a bustling beef cattle industry in the 1880's. Unfortunately the North American fascination with machinery obscured this sound historical economic basis for beef cattle in Alberta and many graziers in the relatively snow-free Chinook region dry-lot their cows and feed hay all winter.

This costly bringing-the-grass-to-the-animal, rather than the opposite, is by no means solely a Canadian phenomenon. I see it even in the Deep South where with just a little management year-round grazing is possible. The root reason for having to feed hay all winter is due to poor autumn grazing management and not building a standing winter feed reserve for winter use.

Today, more and more graziers in Alberta, even those in the non-Chinook snow belt, are finding that intensive grazing management can greatly lower their need for stored forages.

At the Stavely Range Experiment Station, Walter Willms, showed me his work on wintergrazing. Walter has been experimenting with grazing standing corn, winter wheat and stockpiled Rough fescue range. So far, stockpiled Rough fescue is the winner by a long shot.

The Rough fescue is relatively unpalatable in the summer and the cattle naturally avoid it. This allows it to stockpile on its own with lax grazing for winter use. With heavy continuous grazing the range deteriorates to an almost pure Bluegrass stand that has little use as stockpiled feed.

He also showed me some plots of different varieties of Russian wild rye. These grasses will stand above the snow but quickly become hard and strawlike. Walter said they may be more useful as a way of catching snow for moisture than for wintergrazing.

He could not recommend planting corn for wintergrazing in Alberta because the cool summers do not allow the corn to produce enough dry matter for it to be economical. He also is opposed to any plowing in the high wind Chinook belt due to wind-caused soil erosion.

Fencing For Controlling Cattle Position

The graziers I visited in the Chinook range area were using relatively large paddocks to force the cattle into areas they would avoid with continuous grazing such as upland slopes and north facing slopes. Many of the north facing slopes are reverting to willow bush due to lack of grazing pressure.

Without fencing, the cattle tend to concentrate in the valleys and riparian zones along streams. An effort at the Stavely Research Station to lure cattle to graze the steep slopes by placing water there has not been very successful.

Many graziers are concluding that it will take a combination of cattle and sheep to maximize the utilization of the steep foothills and stop the encroachment of yellow poplar and willow on the north slopes.

Norm and Donna Ward of Granum, Alberta, primarily use their limited hay reserves to concentrate cattle on bare spots in their paddocks. Norm has found the concentrated hoof action and dunging to be an excellent way to reseed these bare areas. Several graziers told me they had found that feeding hay in Yellow Poplar groves was also an excellent way to kill out these range invaders.

It appeared that more and more graziers were learning to use hay and feed to create the high stock densities necessary to reseed bare ground and remove brush. They said it was far better to let the cattle spread the manure themselves than to have it heap up in a corral all winter and have to respread it with a machine.

Ward's cows primarily winter on stockpiled Rough fescue and a few pounds of range cake fed every other day. He uses the feed to move the cows to areas where he wants the cows concentrated to create a high stock density. To ease the grazing pressure on his fescue slopes in the summer and allow them to stockpile more winter feed, Norm and Donna now graze some of the alfalfa they formerly cut for hay.

He said it may be that intensively grazing irrigated hay fields in the summer and reserving the upland range primarily for wintergrazing might be the most economical use of both resources particularly when the dramatic lowering of machinery costs this would allow is factored in.

Ward and his chief cowboy, Raymond Benick, drove me to the north face of one of the steeply rolling hills. He showed me the brush encroachment that was starting on all the north facing slopes due to under grazing. He scraped away the thin covering of snow and showed me why it was happening. The ground between the mature clumps of fescue was bare. "That's the start of our desert," he said.

Ward is subdividing his place primarily to force his cows to graze higher on the steep hills and on these undergrazed north faces. "The cows will just stay down in the valley and graze and re-graze the easy land. You've got to force them to graze the slopes," he said. He forces his dry cows to graze off the north faces in winter and remove the old dead material so that new growth can come in during spring.

Later Calving Key to Lowering Costs

Ward's cows start calving at the first of April and he weans them in mid-September. He sends all of his calves to a custom feedlot at weaning and has them fed for 70 days. The "preconditioned calves" are then sold at auction.

Norm has found the primary way to lower wintering costs with beef cows is to calve later in the season. He has moved his calving season back to May to match the start of the green grass season and said it has made all the difference in the world in upping his cows' reproductive rate and cutting the need for hay and cake feeding.

Charlie Ewing, who grazes cows near Claresholm, agreed with Norm. **By moving his calving season back to May and June he has so dramatically cut his hay needs that he has stopped making hay.** He still keeps a hay reserve as an insurance policy against occasional blizzards but finds it is more economical to buy this small amount of hay than to own the equipment necessary to make it.

Ewing's cows graze stockpiled fescue with no feed until late February and fescue and limited supplemental feed after that.

Several graziers commented that the cows would quit working for their feed by grazing if you started haying them too early in the winter. They termed these hay-fed cows "Welfare Cows."

"We've found that cows can bulldoze through as much as two feet of snow to reach the grass underneath, but all that effort stops the day you start to feed hay," Norm Ward said.

As long as the warm Chinook winds keep the snow cover to a minimum, the native grasses cure well as standing hay and cattle can be run with no hay and feed year-round.

"The only problem is that you get about a 60% calving rate doing it that way," rancher Jack Vandervalk of Claresholm warned. "You can't stand that at the price of today's cattle."

Vandervalt supplements his cows with either hay or grain, depending upon the price, from January to the first of May.

However, hay serves as the main feed source only during major weather emergencies. Most of the time the warm Chinook winds keep the rolling bare hills of his ranch blown free of snow and a little protein supplement is adequate for his cows. "We haven't had a big snow here in seven years, now," he recalled when we spoke in 1989.

Vandervalk is a shareholder in the Waldron Grazing Co-op, a 59,000 acre summer grazing ranch. Each shareholder is allowed to graze four cows or eight yearlings for six months per share of stock held. The co-op hires the cowboys necessary to watch members' cattle for the grazing season. This allows cattlemen to fully utilize their land for stored feed production for the winter.

Vandervalk said the high price of land had pushed many ranchers into a stored forage program in the 1970's as a way to increase cattle numbers. He is "interested but skeptical" about the benefits intensive grazing could bring to the short grass areas of Alberta and sees it more as a tool for the soft-stemmed fescue areas.

Vandervalk calves for 50 days starting the third week of March and weans the first week in October. The calves are then wintered on a little hay and grain until May. They are taken to Waldron Co-op and grazed until October and go into the feedlot at 950 to 1000 lbs. to be finished.

Into The Snow Belt

Of course, not all of Alberta is in the Chinook Belt. Most of it lies beneath a foot or more of snow from early November to late March. Still innovative graziers are finding out that snow does not necessarily have to stop the grazing season.

Jim Bauer of the Grey Wooded Forage Association knelt down and brushed the light covering of snow off the lodged over and matted grass. The top layer was yellow and clearly frost damaged, but Bauer reached down into the mat of grass and pulled out a hand full of beautiful green grass. He smiled at my amazement.

"We've had some stockpiled pastures stay as high as 17% protein right through the winter," he said. "The trick seems to be snow cover. Grass that is covered by snow deteriorates far less than that in the open."

The pasture we were examining was a stockpiled winter pasture of Kentucky bluegrass, timothy, fescue, and a legume component of white, red, and Alsike clovers. We were on Don and Randee Halladay's ranch, east of Rocky Mountain House, Alberta, Canada. Rocky Mountain House is some 60 miles north of Calgary in the lee of the Rockies.

The Halladays are among the pioneers in intensive grazing in the area. Don has divided his ranch into 42 permanent paddocks, but has realized that isn't enough to successfully spin out his stockpiled pastures to spring in the long Canadian winter.

"I've got to more tightly ration out these stockpiled pastures and stripgraze them. Right now they enter a paddock and eat the best of it the first couple of days and then they're starving to death," Halladay said.

Unlike most of his neighbors who spend their summers frantically baling hay, Don is purchasing his hay and using his pastures to graze and to build a stockpiled feed wedge that the cattle can winter graze themselves. He runs both cow-calf and yearlings.

Jim Bauer of the forage association said that **cows don't mind rooting through six to eight inches of snow if they know they are going to find something green down there. The stripgrazing technique whereby only a little grass is allocated at a time seems to encourage the snow rooting activity in the cows.**

Cows will readily graze until the snow reaches their eyes. What slows them down is not the snow but the icy crust on the snow from thawing and refreezing. Canadian graziers have discovered that if just a few blades of grass penetrate the snow they will concentrate enough solar heat to prevent an icy crust from forming and will keep the snow soft and pliable all winter long. The trick to wintergrazing in deep snow then is the deepness of the forage mass underneath the snow.

One way to create this is to cut hay fields and swath them into deep windrows in the late fall.

Utilizing swathed oats has been quite successful. This technique consists of planting a crop of oats late in the growing season so that they don't have enough time to mature before frost. At the start of the cold weather the oats are cut and swathed into

windrows just as they would be before being forage chopped or baled. Bauer said oats handled in this manner "sort of freeze-dry" into a green high quality feed that will keep in a cold climate all winter. These windrows are then rationed out to the cows with portable electric fence.

Too many graziers waste this quality by allowing their cattle access to the whole swathed field at once. This not only allows the cattle to select graze a too high quality diet in the early winter with a resulting too low of a quality diet in the late winter, but the cattle will sleep on and step on more feed than they will eat.

"We have found that the upright growing, hay type forages do not stockpile well as standing forages. However, they can make a very high quality winter feed if swathed and left in the field as a deep windrow," explained Bauer. Oats, wheat, or annual ryegrass all make excellent forages for winter swath grazing, he said.

"You want to have the windrows cleaned up before the ground thaws in the spring to prevent bogging," Bauer advised.

Bauer said the willingness of a cow to root through the snow was in direct proportion to the amount of feed that effort would provide her. By swathing and allowing fescue to deep stockpile in the Autumn, a greater reward could be provided for the cow. Many of the Alberta graziers have roundbale unrollers that put the hay in a windrow to reduce waste, better spread the manure and prevent pugging in wet weather. However, this windrow effect gave at least one Alberta grazier pause.

Dennis Wobeser of Lloydminster told me,"I was out feeding the cows last winter and was putting the hay in a nice windrow, when I suddenly realized all that hay had been in a windrow once before. I had windrowed it, baled it, hauled it off the field, stored it, picked it up, hauled it back to the field and put it back in a windrow. What a huge waste of energy!"

He said that swath grazing could be a way to better integrate beef cows and grain production on the Canadian prairies.

Integrating Grass and Grain Production

Wobeser said that small grain production would probably always be a big part of agriculture on the Canadian prairies because it was an efficient way to capture and store the solar energy in the long days of the Canadian summer but that he no longer wanted any part of it due to the high cost of machinery.

"I think a great many Canadians would also like to quit grain production and concentrate solely on grass farming but they need a source of straw for bedding and feed. What we must do is to better integrate beef cows and grain production without having to do both on the same farm," he said.

Dennis said more and more grain farmers were becoming aware of the huge soil drain of selling straw off their farms and were now reluctant to do it. This had tremendously raised the price of straw in the province and had kept many cattlemen in the grain business just for the straw.

He said that by windrowing the straw on the grain farmer's fields and bringing the cows to the straw rather than vice-versa, the straw cold be converted into readily available nutrients via manure and spread over the field by the cows themselves.

"The trick is use the movable fences to force the cows to clean up each windrow before going to the next one. We're not going to make any friends with our grain farming neighbors if all he gets is his straw spread all over the field rather than converted into manure," he said.

160

Olivier La Rocque, a foothill grazier near Pinder Creek told me he now "partnered" with a grain farmer on the prairie. Olivier said the grain farmer grazes his cows on Olivier's foothills ranch in the summer and Olivier in turn grazes his cows on the grain farmer's straw fields in the winter. "It is a six month, six month partnership and sure works better for both of us," he said.

Jack Olsen of Red Deer said that Alberta had a huge resource of cow quality winter feed in stockpiled range and small grain straw. However, the flip side of that was that the province's pastures were probably of too high a quality to waste on beef cows in the growing season.

Importing Calves

"We need more stocker cattle in the summer and more cows in the winter to maximize the profit from our forages," he said.

One way Jack and Dennis have been doing this is to contract graze imported Hawaiian calves for the Parker Ranch.

"We can get two and a half to three pounds a day on these Hawaiian Brangus cattle for a hundred or more days on our summer pastures. The profitability of contract grazing steers sure beats beef cows." he said.

Animal behaviorist Bud Williams is trying to solve that problem by teaching Alberta graziers how to herd their cows in the dense bush that covers the northern half of the province. "With herding we can take the cows up into the bush and use the open pastures exclusively for stocker cattle," Bud said.

Already Alberta is importing calves from the surrounding Canadian provinces and as Canadian and American feeder cattle prices synchronize more closely in the future, no doubt more American calves will start to wend their way north in years to come as well.

While we were in Alberta, Mary and Jack Vandervalk drove us to the "Smashed in Head Buffalo Jump" Provincial park and museum. This is a museum built into the side of a cliff that buffalo were driven over to their deaths by Indians. The museum pointed out that the Great Plains of North America once supported a herd of 60 million buffalo without one drop of hay, grain, or protein supplement.

The paradigm exists. All we have to do is study it and replicate it in our pastures and ranges.

Hopefully, American graziers who now throw up their hands in despair at the first sniff of snow will also find their way north as well. When it comes to winter the Canadians are the North American experts and there is a lot we can all learn about maximizing our resources and minimizing our climatic flat spots from them.

Chapter 36

Working with Nature Works

A visitor asked Mexican rancher, Joe Finan, when was the last time he fed his cattle. He thought a moment.

"In 1952. Yes, I'm sure that was the last year," he replied. "I don't let my animals die from starvation, but I ship 'em if they can't hack it without supplementation."

It is Finan's belief that both man and his animals should bend to nature's plan. His quiet conservation work in the isolated high mountain valley of Columbia in northern Mexico is considered one of the finest in North America. An early convert to Holistic Resource Management, Finan zealously protects the natural cycle of life in his 20-mile-long valley and accepts bears, lions, droughts, floods, and freezes as part of that cycle.

"If you take care of your land during the good years, it will take care of you during the bad years," he said.

He first heard Allan Savory speak in the late 1970's but didn't start building his grazing cells until 1981. "It wasn't that I didn't believe Savory. I didn't believe the electric fences would work." In 1981, he built two 16 paddock cells with an average size paddock of 300 acres. These two cells are usually run as separate units with a fall calving herd in one and a spring calving herd in the other, but they are combined and run as one big cell during drought periods. "I'd never made it through the last big drought without the cells," he said.

The valley at 4250 feet in elevation provides cool summers, but cold winters. Greenup traditionally does not occur until May. Finan said they have had snow all but seven years out of the last 47. Rainfall averages 12 to 13 inches a year.

Herd Of Herefords

To survive and breed with no hay or supplement under these tough conditions, Finan has spent a lifetime developing an equally tough herd of mountain Herefords. These cattle are kept on a strict 55 day breeding season and are divided into fall and spring calving herds.

Finan was an early convert to vertical integration and for 32 years he has grown out all of his calves and sold them to a local packer in Muzquiz. In 1987, however, the prices on the U.S. side were too high to ignore and Finan exported his steer calves to Del Rio, Texas.

Finan's family started putting together land in the valley in the 1890's. The valley had never been settled, nor stocked with cattle because there was no natural water supply. The valley is an enclosed bowl with no natural stream entrance or exit. A 1000 ft. well finally hit water in 1915, and Finan's family moved from Oklahoma to start stocking the valley.

Today, all the water on the ranch still has to be pumped from two wells at the house and piped for six miles to water the cattle. The centered hub cells help cut water reticulation costs and have allowed a 25% increase in stocking rate.

Finan also pointed out that they help to greatly cut labor costs. "Four cowboys can round up our entire herd in a day," he said.

Bill's wife, Nellie, served a fine beef barbecue for us during our visit. The ranch house features large breezy porches and cool polished tile floors. Like most northern Mexico ranches, the ranch

164

has to be completely self-sufficient and generate its own power. Tidy red and white cowboy houses cluster near the main ranch house and planted fruit and nut trees and lush climbing vines give an oasis feel to the headquarters compound.

Because neighboring ranches are also in valleys, an air mile separation of a few miles can take many hours of driving to get to a mountain pass to cross the mountain range. A tunnel through the mountain range cuts driving time to Muzquiz for the Finans, but the private plane still serves as the family "car."

Finan is sparing in his visitor invitations, explaining that he does not like ranchers who "refuse to see."

"Ranchers in Texas are spending a fortune spraying and plowing mesquite, while we are seeing a natural die-off after only six years of practicing HRM. I don't have time to waste on ranchers who can't see what is happening here with their own eyes."

However, even Finan can get surprised by how fast his range in changing. On a tour of the ranch Finan became temporarily disoriented when the jeep track we were following disappeared in the tall grass.

"Hell, if the grass gets much better, I'll have to get a satellite guidance system to find my way around on my own ranch," Finan laughed.

Chapter 37

Work Fit for Man

Cattlemen seeking someone to commiserate with about the cattle business will not enjoy a day with Laurie Lasater, manager of ISA (ee-SAH) Cattle Company in San Angelo, Texas. Contrary to popular talk, Lasater believes the cattle business is about to enter a new era of "professionalism" thanks to changes in genetics, grazing technology, and lending.

"The worst thing that happened to the cattle business was equity financing. Money went to those with equity rather than those with management ability. Today, that's changing. Money is seeking out management and ability rather than equity," he said.

With pride, Lasater points out that Isa Cattle Company runs some 2000 cows on ranches in Texas and Florida but does not own one acre of land. These cows are managed with a total crew of four men and four pickups. In 1985 ISA racked up some two million dollars worth of gross sales.

"One man, one pickup, 500 cows, that's where the cow business is going," Lasater believes. "If a man has the management ability, there's no way to keep him from ranching today. During deflation, assets gravitate toward management."

Lasater points out one of the ranches he has leased near San Angelo as an example of his theory. "Here was a ranch with absentee owners, whose leases were continually going broke every six to eighteen months. We took it over and made a success out of it despite the four-year drought here in West Texas."

Lasater pays a lease of $5 per acre, but releases the hunting rights for $4 an acre giving him an effective lease of $1 an acre. The ranch came complete with corrals, watering systems, and a complete grazing cell system.

"All the capital we need in the cattle business is already here. We don't need any more money attracted to the cattle business. We need more good management."

They Went That-A-Way

An early convert to Allan Savory's holistic theory of ranch management, Lasater sponsored Savory's first school when Savory migrated to America from Africa. "Allan Savory helped me see that you don't make progress by making small ten-degree turns. You make progress by making 180-degree turns and going completely the other way."

Lasater's theory of ranching is based upon numbers and efficiency rather than the finer points of cattle raising. **He is a believer in spring calving and forcing the cowherd to adapt to the environment rather than changing the environment for the cows.** He does not start supplementing his cows (whole cottonseed) until late January and puts up no hay.

Following his dad's philosophy that the cow business is based around the cow and not the steer, Lasater said a breeding herd had to be based around the concept that its main task was finding, identifying, and getting rid of bad cattle, which means an extremely heavy cow and heifer cull.

"The steer is strictly a byproduct to a cowman. His stock in trade is producing cull cows and replacement quality heifers," he said. "Any cowman who is directing the majority of his females to the feedlot is totally missing the boat."

When we talked, Lasater said he was on a 70-day calving

167

season, but his goal is to breed all his cows on one heat cycle. By partnering with a rancher on a fall calving season, he is able to run 60 to 80 cows per bull and has a goal of one bull per 100 cows thanks to spreading the bulls over the two calving seasons. All heifers are bred as yearlings, but the yearlings aren't pushed to breed. "We don't want to push them. Breeding them as yearlings is your first important culling," he said.

While a believer in total vertical integration, Lasater prefers to farm out the post-weaning portion to others and concentrate his management on the cow herd. He admits to being somewhat skeptical about "artificial" grasses compared to native grasses.

"If we take the land cost out of the cattle business, which deflation and lower lease rates are doing, are these artificial grasses really that much better for a cowherd?" he asked.

"Every one of the important American breeds was developed under extremely harsh environments. It seems to me that the 'white fence' seedstock concept is totally wrong. Seedstock cattle need to be raised under the same challenging conditions that commercial cattle are," he said.

One often overlooked aspect of going to controlled grazing is the tremendous lowering of labor costs due to the ease of handling the cows.

"A lot of ranches are starving themselves and their cowboys to death due to poor labor productivity. With one hand per 500 cows, you can afford to pay a good wage and get good help. Of course, it sure takes all the romance out of the cow business. There are no horses and chuckwagons or Wild West stuff and a lot of people are disappointed when they come out here and don't find that," he said.

T-Circle Ranch Division

Matt Brown, head of Texas cattle operations, runs the 500-cow T-Circle Ranch, a division of ISA. The ranch consists of 7,575 acres and was originally divided into two 16-cell hubs. Today, however, Matt runs two cells as one. This allows a much longer rest period between grazings, and rest is the key to higher productivity in the dry West Texas area.

"The more cells there are, the more likely it is there will be a rain before you are back on that cell again," he said.

Matt said they started hooking the cells together during the long recent drought in West Texas. Thanks to the cell they were able to get through the drought with a minimum of destocking.

"With the cell we can run approximately one AUM (1000 lbs. of body weight) per ten acres, compared to one per 16 to 20 acres under continuous grazing," he said.

Matt advises ranchers considering building grazing cells to put lateral gates in the fence near the center of the hub, as well as the hub gate. With the lateral gates, the lateral gate to the next cell could be opened and the hub gate closed and the cattle would naturally cycle to the next cell with a minimum of attention from the cattleman.

The cell fences are two wires set high enough that the calves can get under them and forward creep the next cell before the cows are turned in on it. By doing this, the calves are always getting the best forage before the cows. Weaning weights on the calves average 600 to 700 lbs. with no supplementation.

"You know, I've never met a good rancher who wanted to be a doctor or lawyer," Lasater said, "but I've sure met a lot of doctors and lawyers who want to be ranchers. Like my dad said, 'Ranching is work fit for man.'"

Glossary

AI Artificial insemination.

Aftermath: Forage that is left or grown after a machine harvest such as corn stalks or volunteer wheat or oats. Also called the "Fat of the Land."

Animal unit day: Amount of forage necessary to graze one animal unit (one dry 1100 lbs. beef cow) for one day.

Annual leys: Temporary pastures of annual forage crops such as annual ryegrass, oats or sorghum-sudangrass.

AU Animal unit: One mature, non-lactating cow weighing 500 KGs. (1100 lbs.) or its weight and class equivalent in other species. (Example: 10 dry ewes equal one animal unit.)

AUM Animal unit month: Amount of forage needed to graze one animal unit for a month.

Blaze graze: A very fast rotation used in the spring to prevent the grass from forming a seedhead. Usually used with dairy cattle.

Break grazing: The apportioning of a small piece of a larger paddock with temporary fence for rationing or utilization purposes.

Breaks: An apportionment of a paddock with temporary electric fence. The moving of the forward wire would create a "fresh break" of grass for the animals.

Cell: A grouping of paddock subdivisions used with a particular set or class of animals. During droughts, several cells and their animals may be merged and operated as one very large cell and herd for rationing purposes.

Clamp: A temporary polyethylene covered silage stack made in the pasture without permanent sides or structures.

Composting: The mixing of animal manure with a carbon source under a damp, aerobic environment so as to stabilize and enhance the nutrients in the manure.

Continuous grazing: See below.

Continuous stocking: Allowing the animals access to an entire pasture for a long period without paddock rotation.

Coppice: Young regrowth on a cut tree or bush.

Compensatory Gain: The rapid weight gain experienced by animals when allowed access to plentiful high quality forage

after a period of rationed feed. Animals that are wintered at low rates of gain and are allowed to compensate in the spring frequently weigh almost the same by mid-summer as those managed through the winter at a high rate of gain.
Also known as "pop."

Creep grazing: The allowing of calves to graze ahead of their mothers by keeping the forward paddock wire high enough for the calves to go under but low enough to restrain the cows.

CWT: 100 pounds.

Deferred grazing: The dropping of a paddock from a rotation for use at a later time.

Dirty Fescue: Fescue containing an endophyte which lowers the animal's ability to deal with heat. Fescue without this endophyte is called Fungus-free or Endophyte-free.

Dry matter: Forage after the moisture has been removed.

Easy feed silage: The bringing of silage to the animal with machinery. Opposite of self-feeding.

Flogging: The grazing of a paddock to a very low residual. This is frequently done in the winter to stimulate clover growth the following spring.

Frontal Grazing: An Argentine grazing method whereby the animals grazing speed is determined with the use of a grazing speed governor on a sliding fence.

Grazer: A animal that gathers its food by grazing.

Grazier: A human who manages grazing animals.

Heavy metal: Large machinery.

Holistic Resource Management: A management discipline and thought model that encourages the seeing of the ranch as a unified "whole."
This term is copyrighted.

HRM: Acronym for Holistic Resource Management.

K: Potassium.

Lax grazing: The allowing of the animal to have a high degree of selectivity in their grazing. Lax grazing is used when a very high level of animal performance is desired.

Ley pasture: Temporary pasture. Usually of annuals.

Leader-follower Grazing: The use of a high production class of animal followed by a lower production class. For example, lactating dairy cattle followed by replacements. This type of grazing allows both a high level of animal performance and a high level of pasture utilization. Also, called first-last grazing.

Lodged over: Grass that has grown so tall it has fallen over on itself. Most grasses will self-smother when lodged. A major exception is Tall fescue and for this reason it is a prized grass for autumn stockpiling.

Management-intensive grazing or MiG: The thoughtful use of grazing manipulation to produce a desired agronomic and/or animal result. This may include both rotational and continuous stocking depending upon the season.

Mixed grazing: The use of different species grazing either together or in a sequence.

Mob grazing: A mob is a group of animals. This term is used to indicate a high stock density.

Oklahoma bop: A low stress method of dehorning stocker and feeder cattle whereby a one to two inch stub of horn is allowed to remain. Widely used in the South and Southwest.

P: Phosphorus.

Paddock: A subdivision of a pasture.

Pastureland: Land used primarily for grazing purposes.

Pop: Compensatory gain.

Popping Paddocks: Paddocks of high quality grass and legumes used to maximize compensatory gain in animals before sale or slaughter.

Pugging: Also called bogging. The breaking of the sod's surface by the animals hooves in wet weather. Can be used as a tool for planting new seeds.

Put and take: The adding and subtracting of animals to maintain a desired grass residual and quality. For example, the movement of beef cows from rangeland to keep a rapidly growing tame stocker or dairy pasture from making a seedhead in the spring and thereby losing its quality.

Range: A pasture of native grass plants.

Rational Grazing: Andre' Voisin's term for management-intensive grazing. Rational meant both a thoughtful approach to grazing and a rationing of forage for the animal.

Residue: Forage that remains on the land after a harvest.

Residual: The desired amount of grass to be left in a paddock after grazing. Generally, the higher the grass residue, the higher the animal's rate of gain and milk production.

Rollback: Light cattle usually sell for a higher price than heavier cattle due to their lower body maintenance. The price spread between light and heavy cattle is called the rollback. See also Value of Gain.

Seasonal grazing: Grazing restricted to one season of the year. For example, the use of high mountain pastures in the summer.

Self feeding: Allowing the animals to eat directly from the silage face by means of a rationing electric wire or sliding headgate.

Set stocking: The same as continuous stocking. Small groups of animals are placed in each paddock and not rotated. Frequently used in the spring with beef and sheep to keep rapidly growing pastures under control.

Split-turn: The grazing of two separate groups of animals during one grazing season rather than one. For example, the selling of one set of winter and spring grazed heavy stocker cattle in the early summer and the replacement of them with lighter cattle for the summer and fall.

Spring flush or lush: The period of very rapid growth of cool season grasses in the spring.

Standing hay: The deferment of seasonally excess grass for later use. Standing hay is traditionally dead grass. Living hay is the same technique but with green, growing grass.

Stock density: The number of animals on a given unit of land at any one time. This is traditionally a short-term measurement. This is very difference from stocking rate which is a long term measurement of the whole pasture. For example: 200 steers may have a long-term stocking rate of 200 acres, but may for a half a day all be grazed on a single acre. This acre while being grazed would be said to have a stock density of 200 steers to the acre.

Stocker cattle: Animals being grown on pasture between weaning and final finish. Stocker cattle weights are traditionally from 350 to 850 lbs.

Stocking rate: A measurement of the long-term carrying capacity of a pasture. See stock density.

Stockpiling: The deferment of pasture for use at a later time. Traditionally this is in the autumn. Also known as "autumn saved pasture" or "foggage."

Stripgraze: The use of a frequently moved temporary fence to subdivide a paddock into very small breaks. Most often used to ration grass during winter or droughts.

Swathed oats: The cutting and swathing of oats into large double-size windrows. These windrows are then rationed out to animals during the winter with temporary electric fence. This method of winter feeding is most-often used in cold, dry winter climates.

Transhumance: The moving of animals to an from seasonal range or pasture. For example, the driving of cattle from winter desert range to high mountain summer range.

Value of Gain: The net value of gain after the price rollback of light to heavy cattle has been deducted. To find the net value of gain, the total price of the purchased animal is subtracted from the total price of the sold animal. This price is then divided by the number of cwts. of gain. Profitability is governed by the value of gain rather than the selling price per pound of the cattle.

Wintergraze: Grazing in the winter season. This can be on autumn saved pasture or on specially planted winter annuals such as cereal rye and annual ryegrass.

Index

A
Adams, Bud 6, 91-94
Alabama 5, 83-85, 102, 103
Alfalfa 46-47, 67, 71, 101, 103, 130, 132, 155
Anderson Ranch 6, 95-100
Arkansas 6, 71, 107, 133-135
Arrowleaf 23, 45, 70, 108, 144, 150

B
Bahia 65
Baleage 29-31
Ball, Claudia 105-106
Bauer, Jim 157-159
Beef cattle 7, 9, 13, 19, 20, 39, 62, 63, 71, 74, 87, 105-107, 127, 135, 153
Bermuda 35, 46, 48, 70, 71, 108, 120, 144

Birdsfoot Trefoil 30, 101-103, 132, 142, 147
Bluegrass 28, 67, 130, 132, 154, 158
Bluestem 35
Bulls 14, 23, 33, 34, 42, 54, 58, 63, 70, 75, 79, 92, 99, 110, 118,
          120, 134, 144, 168

C

Canada 153, 158
Cannell, Mike 53, 54
Cereal rye 42, 70, 89, 90, 130, 177
Chickens 5, 15, 17-21, 116, 134, 137, 138
CHTE 149-152
Clover 5, 17, 19, 23, 27, 28, 30, 38, 41-43, 45-46, 61-66, 69-70,
          76, 79, 84, 89, 92, 94, 108, 110, 120, 126, 132,
          141, 144, 150-151, 172
Cockrell, John 51, 52, 54
Collards 137
Cow-calf 6, 26, 33, 56, 62, 63, 66-67, 88, 91, 94, 96, 102, 108,
          141, 143, 145, 158
Crabgrass 23, 38, 63, 84, 120

D

Dairy 23, 30, 50-54, 69-71, 73, 74, 79, 109, 110, 115-116, 118,
          120, 128-129, 135, 171, 173, 174
Dallisgrass 112
Davis, Walt 45-49

E

Eggmobile 20-21
Elizondo, Fernando 131-132

F

Fescue 12, 16-17, 27, 28, 30, 46-49, 66-67, 84-85, 87-89, 108,
          130, 132, 134, 150, 154-159, 172-173
Finan, Joe 163-165
Florida 33, 83, 91-94, 120, 166

G

Gamagrass 135
Gates, George 87-90
Gin trash 40, 43, 44
Goats 28, 105, 106, 126, 129, 134
Grama 112

H

Halladay, Don and Randee 158
Hawaii 6, 74-76, 78, 123, 125, 126, 140, 141, 149, 151
Heifer 6, 15, 63, 68, 109, 133, 134, 136, 139, 167
Heifer Project International 133-139
Hicks, Joseph and Becky 109-110
Hogs See Pigs
Hunting leases 33, 65, 96, 100, 105, 152, 167

I

Indiangrass 112

J

Johnsongrass 46, 47, 84, 98, 112, 117, 120, 138, 143-145, 150-
           151
Josey, Clint 143-145

K

Kahuka Ranch 140
Kaufmann, Diane 114-116
Kentucky 26, 67, 158
Kikuyu 75-77, 124, 126, 140
King Ranch 5, 32-36

L

La Rocque, Olivier 161
Lambs 15, 76, 114, 135-136, 146-147
Lasater, Laurie 166-169
Leader-follower 14, 67, 70, 76, 124, 144, 173
Legume 36, 45, 48, 49, 91, 102, 130, 132, 158
Lespedeza 137

Leys 29-31, 170
Loftin, Gary 95-100
Louisiana 36, 42, 55, 57-60, 71, 92
Ludlow, Reid 101-104

M

McBee, Ron 66-68
McCarville, Paul 52, 53
Meaux, J.B. 57-60
Medics 30
Meucci, Bob 117-122
Mexico 6, 52, 55, 60, 100, 111-113, 128-130, 153, 163-164
Mississippi 22-26, 38-44, 61-65, 117-122
Missouri 44, 66, 71, 87, 89

O

Oats 23, 35, 70, 81, 110, 137, 158-159, 170, 176
Oklahoma 45, 108, 145, 164, 174
Olsen, Jack 161
Opitz, Charles 50, 52, 53
Orchardgrass 12, 17, 89, 103, 110, 126, 130, 132
Osuna, Guillermo 111-113

P

Parker Ranch 6, 78, 123-125, 127, 161
Patenaude, Dan 53
Peanuts 137
Peas 30, 39
Pennsylvania 80
Pigs 5, 38-40, 42, 80-82, 133-135, 137, 139
Plantain 30
Poultry, see Chickens
Pulvermacher, Carl and Kathy 52, 54

R

Rape 137
Replacement heifers 36, 42, 52, 53, 99, 124, 129
Reynolds, Pete, Jr. and Sr. 83-86

Rogers, Harland 22-24
Roth, Steve and Pam 69, 73
Ryegrass 23, 24, 30, 38-39, 63-64, 69-70, 75-76, 79, 84-85, 101-
         102, 108, 110, 130, 132, 134, 144, 150, 159, 170,
         177

S
Sainfroin 132
Salatin, Joel and Teresa 11-21, 116
Sands, Bunker 149-152
Seagrest, Jeff and John 61, 62, 65
Shademobile 13-14
Sheep See Lambs
Smith, Larry 101-104
Snyder, Jodi and Ron 80-81
Sows see Pigs
Spencer, Hub 26-28
Stallings, Alan 107, 108
Steers 15, 23-29, 34, 63, 84, 96, 98-99, 103, 118, 130, 152, 161,
         176
Stocker 7, 8, 22-24, 26, 29, 31, 33, 42, 46-47, 62-63, 66-67, 71,
         74-76, 84-86, 88, 96-99, 101-102, 104, 107-108,
         118, 120, 124, 128-130, 135, 141, 145, 150, 152,
         161, 174-176
Stokes, Norman 83, 85
Sudex 137
Sweet corn 81, 137
Switchgrass 143-145, 150

T
Texas 5, 32, 34-36, 60, 69, 73, 84, 95-100, 105, 111, 129, 143,
         145, 149-153, 164-169
Triticale 120, 121
Turnips 81, 110, 137

V
Vandervalk, Jack 156-157, 161
Verdhoff, Carl 140
Vetch 23, 30, 42

Virginia 5, 11, 12, 17, 18, 29
Von Holt, Pono and Angie 74-76

W

Ward, Norm and Donna 155, 156
Wheat 29, 35, 38-39, 41, 45, 47, 78, 85, 108, 120, 130, 132, 152,
        154, 159, 170
Whitfield, Jim 38-44
Willms, Walter 154
Wintergrazing 5, 18, 40, 41, 55, 56, 88, 120, 147, 154, 155, 158
Wisconsin 50-53, 101-104, 114, 115, 129, 146, 147
Wobeser, Dennis 160
Woods, Peter 146-148

Author's Bio

Allan Nation has been the editor of **Stockman Grass Farmer** magazine since 1977. The son of a commercial cattle rancher, Nation has traveled the world studying and photographing grassland farming systems. As a speaker, he has been a featured presenter in the United States, Canada, Mexico, New Zealand, and Ireland. In 1987, he authored a section on intensive grazing in the <u>USDA Yearbook of Agriculture</u> and has served as a consultant and resource for Audubon Society Television Specials, WTBS, PBS and National Public Radio. He was presented the 1993 Agricultural Conservation Award from the American Farmland Trust for spearheading the drive behind the grass farming revolution in the USA. He is also the author of **Pa$ture Profit$ with Stocker Cattle,** and **Quality Pasture**. Allan is married to novelist Carolyn Thornton and they live in Mississippi.

For more books by Allan Nation, please turn the page....

Pa$ture Profit$ With Stocker Cattle
by Allan Nation

America's first book on stocker grazing is written for those who want to get rich with a minimum of financial risk.

In **Pa$ture Profit$ With Stocker Cattle,** Allan Nation, editor of the **Stockman Grass Farmer,** illustrates his economic theories on stocker cattle by profiling Mississippi grazier, Gordon Hazard. Famous in national beef cattle circles for his penny-pinching ways, Hazard claims never to have lost a dime on stocker cattle in over 40 years of graziering. **Pa$ture Profit$ With Stocker Cattle** shows how Hazard has accumulated and stocked a 3000-acre ranch solely from retained stocker profits with no bank leverage.

"Truly outstanding, something no cattleman should be without--at least if he depends on grass for his sustenance." **Livestock Weekly**
"Filled with good information on running a stocker cattle business, the book provides information that can be applied to improve business management techniques for other businesses as well." **Small Farm Today**
"(Nation) takes an altogether different view of stock raising, dealing with bankers, marketing, investment in equipment, etc., than most of us grew up with...If this book doesn't create controversy and spark some soul searching, particularly in the cattle business, both beef and dairy, it isn't for lack of effort on the part of the author." **Draft Horse Journal**

Pa$ture Profit$ With Stocker Cattle
ISBN: 0-9632460-0-3
Softcover 192 pages $24.95 + shipping and handling

Grass Farmers
by Allan Nation

If you're tired of reading farm stories of doom and gloom, bankruptcies and despair, then **Grass Farmers** will tell you about the many people who are making an excellent living on the land.

37 thought-provoking success stories tell
- how sheep dairying can produce a quality life from small acreages,
- how a grazier paid for his farm in six months,
- how to create a grass farm as a retirement job,
- how heifer grazing can give a grazier a mid-winter vacation,
- and much more, including a grazier's glossary.

Allan Nation has traveled the world studying and photographing grassland farming systems, profiled in this collection. Many of these stories appeared in now out-of-print editions of the **Stockman Grass Farmer**, where Nation has been editor since 1977.

"The book won't tell you how big to make your paddocks or when to move your stock. But it is a quick and inspirational read that could put many graziers on a path to higher profits." **New Farm**

"We guarantee it will get you to thinking about new ways to increase your profit as a Grass Farmer." **Livestock Market Digest**

"Interesting to see another face to farming. Great dream material." **New England Farm Bulletin**

Grass Farmers
ISBN: 0-9632460-1-1
Softcover, 192 pages. $23.50 + shipping and handling

Quality Pasture, How to create it, manage it, and profit from it
by Allan Nation

Do you know how to cut your current feed costs in half?

Quality Pasture, How to create it, manage it, and profit from it offers down-to-earth, low-cost tactics to create high-energy pasture that will reduce or eliminate expensive inputs or purchased feeds. **Quality Pasture** is the first book of its kind directed solely toward farmers like you who are beginning or practicing management-intensive grazing with ruminant livestock. Chapters cover:
- extending the grazing season during winter and summer slumps,
- matching stocking rates with pasture growth rate,
- how to create a drought management plan,
- tips for wet weather grazing,
- and a detailed section on making pasture silage.

Quality Pasture will walk you through the production model that can help you plug your profit leaks. Examples of real people making real profits show that quality pasture is not only possible, but can be profitable for you, too. Chapter summaries give you plenty of food for thought and action tips that you can begin using now.

"Among the people I admire most in agriculture are those who advocate new ideas and practices with unwavering conviction, passion and strength of will....Allan Nation speaks with the same passion he puts into his writing." **Successful Farming**

Quality Pasture
ISBN: 0-9632460-3-8
Softcover, 288 pages. $32.50 + shipping and handling

For a complete list of Green Park Press books and SGF Special Reports, request our FREE catalog.
Call 1-800-748-9808

GREEN PARK PRESS

THE STOCKMAN GrassFarmer

Green Park Press books and the Stockman Grass Farmer magazine are devoted solely to the art and science of turning pastureland into profits through the use of animals as nature's harvesters. To order a free sample copy of the magazine, or to purchase other Green Park Press titles:

Please make checks payable to:

Stockman Grass Farmer
PO Box 9607
Jackson, MS 39286-9607

1-800-748-9808
or 601-981-4805
FAX 601-981-8558

Shipping:	Amount	Canada	Mexico
Under 2 lbs.	$3.50	$5.50	$7.50
2-3 lbs.	$4.75	$7.50	$12.00
3-4 lbs.	$5.25	$8.50	$15.00
4-5 lbs.	$6.50	$10.00	$16.75
5-6 lbs.	$8.50	$11.75	$21.00
6-8 lbs.	$12.00	$13.50	$24.00
8-10 lbs.	$14.50	$16.75	$30.00

Foreign Postage: Add 35% of order.

- -

Name _____

Address _____

City _____

State/Province_____ Zip/Postal Code_____

MC/VISA # _____ Expiration _____

Signature _____

Quantity	Title	Price Each	Sub Total
_____	Grass Farmers (weight 1 lb.)	$23.50	_____
_____	Quality Pasture (weight 1 1/2 lbs.)	$32.50	_____
_____	Pa$ture Profit$ With Stocker Cattle (1 lb.)	$24.95	_____
_____	Free sample copy Stockman Grass Farmer magazine		

Sub Total _____

Postage & Handling _____

Mississippi Residents Add 7% Sales Tax _____

U.S. Funds Only, Please   TOTAL: _____

Name _____

Address _____

City _____

State/Province _____ Zip/PostalCode _____

MC/VISA #_____ Expiration _____

Signature _____

Quantity	Title	Price Each	Sub Total
_____	Grass Farmers (weight 1 lb.)	$23.50	_____
_____	Quality Pasture (weight 1 1/2 lbs.)	$32.50	_____
_____	Pa$ture Profit$ With Stocker Cattle (1 lb.)	$24.95	_____
_____	Free sample copy Stockman Grass Farmer magazine		
		Sub Total	_____
		Postage & Handling	_____
	Mississippi Residents Add 7% Sales Tax		_____
U.S. Funds Only, Please		**TOTAL:**	_____

- -

Shipping:	Amount	Canada	Mexico
Under 2 lbs.	$3.50	$5.50	$7.50
2-3 lbs.	$4.75	$7.50	$12.00
3-4 lbs.	$5.25	$8.50	$15.00
4-5 lbs.	$6.50	$10.00	$16.75
5-6 lbs.	$8.50	$11.75	$21.00
6-8 lbs.	$12.00	$13.50	$24.00
8-10 lbs.	$14.50	$16.75	$30.00

Foreign Postage: Add 35% of order.

Please make checks payable to:
Stockman Grass Farmer
PO Box 9607
Jackson, MS 39286-9607

1-800-748-9808
or 601-981-4805
FAX 601-981-8558

Green Park Press books and the **Stockman Grass Farmer** magazine are devoted solely to the art and science of turning pastureland into profits through the use of animals as nature's harvesters.

STOCKMAN GRASS FARMER
PO BOX 2300
RIDGELAND MS 39158
1-800-748-9808

THE STOCKMAN Grass Farmer